Mathematica® Computer Programs for Physical Chemistry

Springer

New York
Berlin
Heidelberg
Barcelona
Budapest
Hong Kong
London
Milan
Paris
Santa Clara
Singapore
Tokyo

William H. Cropper

Mathematica®
Computer Programs
for Physical Chemistry

With 11 Illustrations and a CD-ROM

Springer

William H. Cropper
Department of Chemistry
St. Lawrence University
Canton, NY 13617, USA

Cover illustration: Three-dimensional plot of a potential-energy surface for a chemical reaction of the kind $A + B \rightarrow AB + C$ using the program Leps described in Chapter 11. The calculation method is that of London, Eyring, Polanyi, and Sato.

Library of Congress Cataloging-in-Publication Data
Cropper, William H.
    Mathematica® Computer programs for physical chemistry / William H. Cropper.
        p.   cm.
    Includes index.
    ISBN 0-387-98337-6 (softcover : alk. paper)
    1. Chemistry, Physical and theoretical—Computer programs.
    I. Title.
    QD455.3.E4C76   1998
    541.3'0285'53—dc21   97-34136

Printed on acid-free paper.

Production coordinated by Chernow Editorial Services, Inc., and managed by Francine McNeill; manufacturing supervised by Joe Quatela.
Typeset by Asco Trade Typesetting Ltd., Hong Kong.
Printed and bound by R.R. Donnelley and Sons, Harrisonburg, VA.
Printed in the United States of America.

9 8 7 6 5 4 3 2 1

ISBN 0-387-98337-6 Springer-Verlag New York Berlin Heidelberg   SPIN 10645072

Preface

Physical chemistry is a science of theories, concepts, and lengthy calculations. To the student the theories and concepts are likely to be more inspiring than the calculations, but that need no longer be the case. With the advent of computer hardware and software, the tedium of calculational work has largely been overcome. Complicated calculations are now accessible to the student through computer programs, and they can become as much a part of the learning experience as the theories and concepts.

This book acquaints you with that pleasant state of affairs by introducing more than 140 computer programs that perform the essential calculations of physical chemistry efficiently and painlessly. Calculational methods covered include those for integrating systems of differential equations, solving systems of nondifferential equations, data fitting, two- and three-dimensional plotting, contour plotting, matrix manipulations, data-file manipulations, Fourier transforms, simulations, and miscellaneous number crunching.

Topics in the book are organized to follow the order found in most contemporary physical chemistry texts. Since the book is not itself a text, general coverage of standard topics is limited to brief summaries and reminders. If that is not enough for you, consult your favorite text. A few topics are specialized and perhaps not familiar; they are given more extended treatments.

Most of the programs are written in *Mathematica*® code. If you are not familiar with the *Mathematica*® language, you will need to spend a few hours (no more) with one of the several introductory books (e.g., T.W. Gray and J. Glynn, *Beginner's Guide to Mathematica, Version* 3, Cambridge University Press, Cambridge, 1997). The programs include instructions on how to enter data and comments on how to interpret the code at each stage of the calculation. Pay attention to these comments: they can save you much wasted time.

For a preview of the programs, you can browse the Contents, where the programs are listed in the contexts of the chapters, and in Appendix A, where they are listed alphabetically with one-sentence descriptions. Two versions of each program are included on the disc that accompanies the book, one (with the extension `.ma`) is prepared for *Mathematica*®, Version 2, and the other (with the extension `.nb`) for *Mathematica*®, Version 3. The disc also includes files for most of the exercises and examples in the book. These files begin with `Ex` (for exercises) or `Exm` (for examples). Data files on the disc have the extension `.m` or `.dat`. The programs are all written in the "one-dimensional" form (i.e., all of the characters on one line). If you want to see the corresponding "two-dimensional" form, open a `.ma` file in *Mathematica*®, Version 3, and request that it be converted with the input in `StandardForm`.

Canton, New York   William H. Cropper

Contents

*Items in mono type are program names.

How to Run the Programs

The CD-ROM disc that accompanies this book contains the programs, applications (in exercises and examples), and required data files. You can run the programs directly from the disc or copy the files of interest to a directory on a hard disk. In any case, at the beginning of a *Mathematica*® session you should first set the current working directory to the one you are using. For example, if you are running the programs directly from the CD on drive E, initially type the command `SetDirectory["E:"]`. This ensures that programs can open data files from the disc. Access programs by selecting Open under the File menu and double clicking the name of the program.

The command `ClearAll["Global`*"]`, included on the first or second line of each program, is intended to clear all entities created in previous programs that might confuse the present program. Occasionally this precaution does not work, and you will have to take more drastic measures, such as turning the *Mathematica*® kernel off and then back on again.

Five of the programs are written in BASIC code, and they run under QuickBASIC, Version 4.5 (*Mathematica*® is too slow for the Monte Carlo and molecular dynamics calculations performed by these programs). All of the BASIC programs have graphics features requiring data from the file GRAPH.DAT. To create this file, run the BASIC program PIXEL.BAS, and enter the graphics mode you are using (choices are CGA, EGA, and VGA). Three of the programs, MC1.BAS, MC2.BAS, and MD3.BAS, also require the subprogram PLOTAXES.BAS. To run one of these programs, MC1.BAS, for example, *open* MC1.BAS first and then *load* PLOTAXES.BAS.

1
Preliminaries

This book is about numbers, so it is fitting that this first chapter opens with a section on numbers, emphasizing units and uncertainties. The computer program Error, introduced in Sec. 1.1, assesses the propagation of uncertainty in ordinary calculations. Another major theme in the book is equation solving. Some aspects of that topic are developed in Sec. 1.2 and applied to a selection of gas laws, ranging from the simplest to one of the most complex. The programs presented in Sec. 1.2, in order of increasing complexity and named after the inventors of the gas laws, are Waals, Peng1, Beattie, Anderko, and Keenan1. Also mentioned in Sec. 1.2 are gas laws in the generic virial form and two programs, Virial1 and Virial2, illustrate. A third theme, which will be important throughout the book, is the interpretation of experimental data by fitting data sets to empirical equations. In Sec. 1.3 the technique of least-squares (or regression) analysis is applied to this task, and the program Linreg is introduced.

1.1 Uncertainties and Units

As an arithmetic problem, the statement $2 \times 3 = ?$ is complete with the answer 6. As a statement involving physical quantities, however, it is incomplete. To find the answer to the corresponding physical problem, you need to know uncertainties and units for the multiplied factors and also for the product.

Let both factors have cm units. Since units behave like algebraic quantities, the product has the units $cm \times cm = cm^2$. Suppose the factors with their uncertainties included are 2.0 ± 0.1 and 3.0 ± 0.2. What is the uncertainty of the product? That question is answered by the *error-propagation formula* (Hecht, 1990, p. 286), which is implemented in the program Error.

Error states the product with the function calcFunction (in the fourth line of code, after the long comment concerning special constants). The two factors are represented symbolically with x1 and x2 (or with

1

any other convenient symbols), in `calcFunction` and also in the list `variables`. Numerical values and uncertainties for the factors are supplied in the list `data`. The program calculates 6.0 ± 0.5. The answer to the original problem, including units, is 6.0 ± 0.5 cm^2.

Here are two more examples in which `Error` is applied.

Example 1-1. Calculate $1.68/(6.325 + 4.56)$, including the uncertainty, assuming that the numbers involved have an uncertainty of one in the last digit.

Answer. See the file `Exm1-1` for details on how to formulate `calc-Function` and `data` for this calculation. The program calculates 0.13038 ± 0.0008.

Example 1-2. The number of Joule energy units in an "atomic unit" of energy is calculated with

$$\frac{m_e e^4}{4\varepsilon_0^2 h^2},$$

in which m_e and e are the mass and charge of an electron, ε_0 is the vacuum permittivity and h is Planck's constant. Calculate this quantity.

Answer. All of the numbers involved in this calculation are "special constants," whose values are supplied in the program `Error`, and uncertainties are small enough to be neglected in our calculations. Thus `calcFunction` is defined with the symbols noted in the program for the special constants and `variables` and `data` are defined as empty lists. See the file `Exm1-2` for details. The program calculates 4.3597×10^{-18} J.

1.2   Gas Laws

Ideal Gas Law

This is where classical and statistical thermodynamics begin. The ideal gas law was one of the first, and is still one of the most important, mathematical models in physics and chemistry. In classical thermodynamics it is stated

$$PV = nRT,$$

in which $R = 0.083145$ L bar K^{-1} mol^{-1} is the *gas constant*, and the most convenient units for the pressure P, volume V, molar amount n, and absolute temperature T are bars (bar), liters (L), moles (mol), and Kelvins (K). In later problems, the gas constant will be expressed in other units, $R = 8.3145$ J K^{-1} mol^{-1}. When both values of R are needed in a calculation (rarely necessary), use $R1$ for 8.3145 J K^{-1} mol^{-1} and $R2$ for 0.083145 L bar K^{-1} mol^{-1}. In statistical thermodynamics the ideal gas law is expressed

$$PV = Nk_{\mathrm{B}}T,$$

where $k_{\mathrm{B}} = 1.3807 \times 10^{-23}$ J K^{-1} is *Boltzmann's constant* and N is the number of molecules.

Nonideal Gas Laws

For an ideal model, the ideal gas law is remarkably realistic. It is accurate for many purposes at pressures below about 10 bar and at temperatures above the boiling point of the component involved. But it may fail under more extreme conditions, and that limitation has inspired the invention of numerous nonideal gas laws. Here are a few of them.

Virial Equations

Nonideal gas laws tend to be mathematically diverse, but many can be rearranged into a generic form called a *virial equation*, usually written

$$Z = 1 + \frac{B}{V_{\mathrm{m}}} + \frac{C}{V_{\mathrm{m}}^2} + \frac{D}{V_{\mathrm{m}}^3} + \cdots, \tag{1.1}$$

in which $V_{\mathrm{m}}\ (= V/n)$ is the *molar volume*, $Z\ (= PV_{\mathrm{m}}/RT)$ is the *compressibility factor*, and B, C, D, ... are second, third, fourth, ... *virial coefficients*, which depend on temperature but not on pressure, although the pressure range covered can dictate how many coefficients are needed.

Most applications of the virial equations utilize just one or two of the virial coefficients. If only B is needed, the equation is

$$Z = 1 + \frac{B}{V_{\mathrm{m}}}. \tag{1.2}$$

If two virial coefficients, both B and C, are required, the virial equation is

$$Z = 1 + \frac{B}{V_{\mathrm{m}}} + \frac{C}{V_{\mathrm{m}}^2}. \tag{1.3}$$

The programs `Virial1` and `Virial2` solve the last two equations and calculate nonideal molar volumes. The next example illustrates.

Example 1-3. Use `Virial2` to calculate the molar volume of benzene at 25 bar and 300 °C. Virial coefficients (from Dymond and Smith, 1969, p. 127) are $B = -0.315\ \mathrm{L\,mol^{-1}}$ and $C = 36800\ \mathrm{L^2\,mol^{-1}}$ at 300 °C and pressures as high as 25 atm.

Answer. Enter data for P, T, B, and C as instructed at the beginning of the program. Run the program and calculate $V_m = 1.5476\ \mathrm{L\,mol^{-1}}$ compared to $V_m = 1.9063\ \mathrm{L\,mol^{-1}}$ for an ideal gas under the same conditions.

---

van der Waals Equation

This is the grandfather of the nonideal gas laws. It was introduced in 1873 by J.D. van der Waals in his doctoral dissertation. Like the virial equation (1.3), it has two parameters (a and b). It is usually written to resemble the form of the ideal gas law,

$$\left(P + \frac{a}{V_m^2}\right)(V_m - b) = RT, \tag{1.4}$$

or as a calculation of pressure,

$$P = RT/(V_m - b) - a/V_m^2. \tag{1.5}$$

Values of the van der Waals parameters for selected gases are listed in Table 1.1, and the same data are included in the file `Chap1.m`, which is called by all of the programs that implement nonideal gas laws. The program `Waals` calculates molar volumes by solving Eq. (1.4) for a selected gas and a given pressure and temperature.

TABLE 1.1. Van der Waals parameters for selected gases

Gas	$a/(\mathrm{bar\,L^2\,mol^{-2}})$	$b/(\mathrm{L\,mol^{-1}})$
He	0.03412	0.02370
Ne	0.2107	0.01709
O_2	1.360	0.03183
N_2	1.390	0.03913
CH_4	2.253	0.04278
HCl	3.667	0.04081
CO_2	3.592	0.04267
NH_3	4.170	0.03707
H_2O	5.464	0.03049
Cl_2	6.493	0.05622
Hg	8.093	0.01696

Source: *CRC Handbook of Chemistry and Physics.*

Example 1-4. Use `Waals` to calculate the molar volume of steam at 200 bar and 500 °C.

Answer. Enter the given pressure and temperature and H2O for the list gas, as instructed in the program. The program calculates $V_m = 0.25880$ mol L^{-1}, compared to $V_m = 0.32144$ mol L^{-1} for an ideal gas. The program also plots an isotherm (constant-temperature) curve for the specified temperature. Run the program for $P = 200$ bar and $T = 600$ K. The `FindRoot` function fails in this case because subcritical conditions are involved, as revealed by the plot of the isotherm. The program starts to search for a root with the ideal gas molar volume, which is much larger than the *liquid* molar volume defined by the van der Waals equation under these subcritical conditions, and `FindRoot` fails to find a root. If the pressure is decreased to 170 bar, however, the program succeeds. Watch out for this problem when you run any of the programs that make nonideal gas law calculations.

Peng–Robinson Equation

This is one of the many elaborations of the van der Waals equation. It can be expressed as

$$\left[P + \frac{a}{V_m(V_m + b) + b(V_m - b)}\right](V_m - b) = RT, \qquad (1.6)$$

or as an equation for calculating the pressure from the molar volume,

$$P = \frac{RT}{(V_m - b)} - \frac{a}{V_m(V_m + b) + b(V_m - b)}, \qquad (1.7)$$

in which a and b, representing another pair of constants (*not* the same as the a and b in the van der Waals equation), are determined by three more fundamental parameters, the critical pressure P_c, the critical temperature T_c, and a much-used parameter ω, called the *acentric factor*, which is a general measure of the geometric complexity of the molecules in the gas. The empirical equations that determine the Peng–Robinson a and b in terms of P_c, T_c, and ω are

$$\kappa = 0.37464 + 1.5422\omega - 0.26992\omega^2$$

$$\alpha = \{1 + \kappa[1 - (T/T_c)^{1/2}]\}^2$$

$$a = 0.4572R^2T_c^2\alpha/P_c$$

$$b = 0.07780RT_c/P_c.$$

A listing of P_c, T_c, and ω values for selected gases is given in Table 1.2.

TABLE 1.2. Critical constants and acentric factors for selected gases

Gas	P_c/bar	T_c/K	ω	$V_{mc}/(\text{L mol}^{-1})$
He	2.27	5.19	−0.365	0.0574
Ne	27.6	44.4	−0.029	0.0416
O_2	50.4	154.6	0.025	0.0734
N_2	33.9	126.2	0.039	0.0898
CH_4	46.0	190.4	0.011	0.0992
HCl	83.1	324.7	0.133	0.0809
CO_2	73.8	304.1	0.239	0.0939
NH_3	113.5	405.5	0.250	0.0725
H_2O	221.2	647.3	0.344	0.0571
Cl_2	79.8	416.9	0.090	0.1238
Hg	1510.	1765.	−0.167	0.0427

Source: R.C. Reid, J.M. Prausnitz, and B.E. Poling, 1987, Appendix A.

These data are included as items 3, 4, 5, and 6 for each gas in the file Chap1.m. The program Peng1 calculates a molar volume according to Eq. (1.7) for a selected gas and a given pressure and temperature; it also plots an isotherm for the temperature chosen. The strategy of the program is identical to that for the program Waals. You can run the program for CO_2 in Exercise 1-6 and compare with similar calculations made with other nonideal gas laws.

Beattie–Bridgeman Equation

The van der Waals equation has two parameters (a and b), and the Peng–Robinson equation has three (P_c, T_c and ω). One route to greater accuracy is to add more parameters. The equation introduced by Beattie and Bridgeman,

$$(P + A/V_m^2)\left[\frac{V_m^2}{(V_m + B)(1 - \varepsilon)}\right] = RT, \qquad (1.8)$$

where

$$A = A_0(1 - a/V_m)$$
$$B = B_0(1 - b/V_m)$$
$$\varepsilon = c/V_m T^3,$$

is defined by five parameters, a, b, c, A_0, and B_0. Values of these parameters are given in Table 1.3 for a few gases and are included as items 7, 8, 9, 10, and 11 for those gases in the file Chap1.m. The program Beattie calculates V_m for a selected gas and at a given pressure and temperature using the Beattie–Bridgeman Equation (1.8), and also plots an isotherm for the

TABLE 1.3. Beattie–Bridgeman parameters for selected gases

Gas	A_0 (bar L^2 mol^{-1})	a (L mol^{-1})	B_0 (L mol^{-1})	b (L mol^{-1})	c (10^4 L K^3 mol^{-1})
He	0.0219	0.05984	0.01400	0.0	0.004
Ne	0.2153	0.02196	0.02060	0.0	0.101
N_2	1.362	0.02617	0.05046	−0.00691	4.20
O_2	1.511	0.02562	0.04624	0.004208	4.80
CO_2	5.0728	0.07132	0.10476	0.07325	66.00
CH_4	2.3071	0.01855	0.05587	−0.01587	12.83

Source: J.A. Beattie and D.C. Bridgeman, 1928.

given temperature. You can run the program in Exercise 1-8 and compare the result with those of two other nonideal gas calculations.

Anderko–Pitzer Equation

This is a more recent (1990) addition to the list of nonideal gas laws. It has the advantage that it achieves good accuracy with only three parameters—P_c, T_c, and ω. It calculates the *molar density* ρ ($= n/V = 1/V_m$) by first calculating the *reduced molar density* $\rho_r = \rho/\rho_c$, where ρ_c ($= 1/V_{mc}$) is the *critical molar density*. The Anderko–Pitzer equation for ρ_r is

$$Z = \frac{1 + c\rho_r}{1 - b\rho_r} + \alpha\rho_r + \beta\rho_r^2 + \gamma\rho_r^3, \tag{1.9}$$

in which b, c, α, β, and γ are empirical parameters which depend on the acentric factor ω and the *reduced temperature* T_r ($= T/T_c$). The program Anderko calculates molar volumes with Eq. (1.9) for a selected gas at a given pressure and temperature, and like the other programs mentioned in this section, also plots an isotherm for the given temperature. For some practice with Anderko see Exercise 1-8.

Keenan–Keyes–Hill–Moore Equation

This may be the world's most complicated gas law. It fits data available on steam and liquid water with 55 parameters. It is one of a set of equations derived to calculate all of the thermodynamic properties of water. (We will see how that is done in Chapter 2). The program Keenan1 implements the Keenan–Keyes–Hill–Moore equation at a selected pressure and temperature, and you can consult the program for more on how the calculation is formulated.

Comparisons

Here is a comparison of ideal and nonideal calculations made with some of the programs mentioned: all calculations are for steam at 200 bar and 500 °C.

Gas law	Program	$V_m/(\text{L mol}^{-1})$
Ideal		0.3214
van der Waals	Waals	0.2588
Peng–Robinson	Peng1	0.2617
Anderko–Pitzer	Anderko	0.2664
Keenan–Keyes–Hill–Moore	Keenan1	0.2661
Experimental		0.2661

1.3   Data Fitting

Suppose you have measured the molar heat capacity C_{Pm} of graphite over a wide range of temperatures and obtained the data points plotted in Figure 1.1. Your aim is to express C_{Pm} as a function of the temperature T. This is a job for the *least squares* method of data fitting.

Two kinds of temperature *fitting functions* are commonly used to fit molar heat capacity (and other) data. The most versatile are polynomial functions

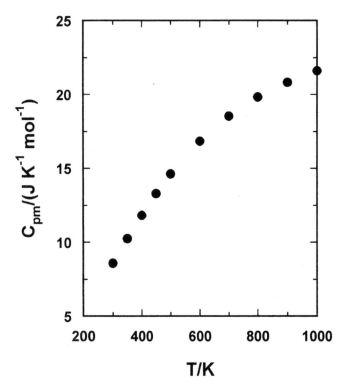

FIGURE 1.1. Molar heat capacity data for graphite at high temperatures.

having the form

$$C_{Pm}(T) = c_0 + c_1 T + c_2 T^2 + \cdots, \tag{1.10}$$

in which $c_0, c_1, c_2, \ldots$ are temperature-independent constants. The number of terms and the maximum power of T needed depend on the temperature range covered: high-temperature data require fewer terms than low-temperature data. The second heat capacity formula in common use is the *Maier–Kelley equation*

$$C_{Pm}(T) = a + bT + c/T^2, \tag{1.11}$$

with a, b, and c the temperature-independent constants. This formula is applicable only to data taken above room temperature.

To fit the graphite heat capacity data we choose a formula of the polynomial or Maier–Kelley kind and adjust values of the constants in the function to make the overall agreement between measured heat capacities and values calculated by the function as close as possible. If $C_{Pm}(T_1), C_{Pm}(T_2), \ldots$ are measured heat capacities at the temperatures $T_1, T_2, \ldots$, and $C'_{Pm}(T_1), C'_{Pm}(T_2), \ldots$ corresponding values calculated by the temperature function (1.10), the object is to minimize magnitudes of the differences

$$C_{Pm}(T_1) - C'_{Pm}(T_1), C_{Pm}(T_2) - C'_{Pm}(T_2), \ldots.$$

In the standard mathematical procedure *squares* of the differences are minimized; hence the name *least squares* for the method. Another term for this general method, originated by statisticians, is *regression analysis*, or more specifically in the cases we will consider *linear regression*, because the coefficients [the c's in Eq. (1.10) and a, b, and c in Eq. (1.11)] occur linearly in the fitting functions.

When data fitting of this kind is done, it is usually assumed that the independent variable is known exactly and uncertainties in the dependent variable are calculated. Data points plotted in Figure 1.1, for example, represent values of the independent variable T and dependent variable C_{Pm}. The first point is determined by $T = 300$ K and $C_{Pm} = 8.581$ J K^{-1} mol^{-1}.

We will not explore further the mathematical basis of the least-squares analysis. A program called `Linreg`, which performs least-squares data-fitting tasks, is supplied, however, and it is easily used with little knowledge of the mathematical background. Here are two examples in which `Linreg` is applied.

Example 1-5. Use the program `Linreg` to fit the high-temperature graphite heat capacity data, plotted in Figure 1.1 and quoted below, to a cubic function of temperature,

$$C_{Pm}(T) = c_0 + c_1 T + c_2 T^2 + c_3 T^3. \tag{1.12}$$

T/K	$C_{Pm}/(\text{J K}^{-1}\,\text{mol}^{-1})$
300	8.581
350	10.241
400	11.817
450	13.289
500	14.623
600	16.844
700	18.537
800	19.827
900	20.824
1000	21.610

Answer. Enter data pairs in the list data beginning on the third line of Linreg code as instructed in the program. Then, on the line of code following the data list, specify the form of the fitting function with

```
fittingFunction = {1, x, x^2, x^3};
```

The four items in brackets correspond to the four terms in the polynomial temperature function (1.12) with x representing T. Calculate

$$c_0 = -5.799 \pm 3.84 \times 10^{-1}\ \text{J K}^{-1}\,\text{mol}^{-1}$$

$$c_1 = 5.993 \times 10^{-2} \pm 2.02 \times 10^{-3}\ \text{J K}^{-2}\,\text{mol}^{-1}$$

$$c_2 = -4.399 \times 10^{-5} \pm 3.30 \times 10^{-6}\ \text{J K}^{-3}\,\text{mol}^{-1}$$

$$c_3 = 1.145 \times 10^{-8} \pm 1.69 \times 10^{-9}\ \text{J K}^{-4}\,\text{mol}^{-1},$$

and 0.0443 for the *standard deviation* of the fit, an indication of how well the overall data-fitting task has been done. The lower the standard deviation the better the calculated polynomial function fits the data. (The uncertainties indicated for c_0, c_1, c_2, and c_3 are also standard deivations.) Notice that the program plots both the original data pairs as points and a curve calculated according to the fitting function (1.12). Also run Linreg for various other polynomial functions (e.g., with 0, 1, 2, and 4 as the maximum degree of the polynomial) and note the standard deviation and appearance of the plot in each case.

Example 1-6. Use Linreg to fit the above graphite heat capacity data to the Maier–Kelley formula, Eq. (1.11).

Answer. In this case the fitting function is specified with

```
fittingFunction = {1, x, x^-2};
```

and the program calculates

$$c_0 = 1.155 \times 10^1 \pm 1.07 \, \text{J} \, \text{K}^{-1} \, \text{mol}^{-1}$$

$$c_1 = 1.112 \times 10^{-2} \pm 1.19 \times 10^{-3} \, \text{J} \, \text{K}^{-2} \, \text{mol}^{-1}$$

$$c_2 = -6.047 \times 10^5 \pm 8.60 \times 10^4 \, \text{J} \, \text{K} \, \text{mol}^{-1}.$$

As the standard deviations show, the fit in this case is not as good as it was in the last example, with the cubic temperature function, Eq. (1.12). That is expected since the cubic fit was done with four parameters and the Maier-Kelley with three.

1.4   Exercises

Uncertainties

1-1   Calculate $(1.234) \ln(6 \times 10^{23})$, including the uncertainty, assuming that the numbers involved have an uncertainty of one in the last digit (e.g., 1.234 ± 0.001). Use the program Error.

1-2   Calculate $(1.234)e^{10}$, including the uncertainty, assuming that the numbers involved have an uncertainty of one in the last digit. Use the program Error.

1-3   Use the program Error to calculate

$$\frac{\ln(1823) + e^{0.525} + 0.0934}{6.52 + 1.8945 + 10.22},$$

including the uncertainty, assuming that the numbers involved have an uncertainty of one in the last digit.

Nonideal Gas Laws

1-4   Use the program Waals to calculate the molar volume of CO_2 gas at 400 K and 1.00, 10.0, and 100 bar. Compare these van der Waals calculations with ideal calculations. Use data from Table 1.1.

1-5   Use the program Virial2 to calculate the molar volume of CO_2 at 100 bar and 398.17 K. Virial coefficients for CO_2 at 398.17 K and pressures up to 240 atm (from Dymond and Smith, 1969, p. 36) are $B = -62.20 \, \text{cm}^3 \, \text{mol}^{-1}$ and $C = 3623 \, \text{cm}^6 \, \text{mol}^{-2}$. Also make the same calculation using Peng1, Beattie, and Anderko in the next three exercises and compare the results.

1-6   Use the program `Peng1` to calculate the molar volume of CO_2 at 100 bar and 398.17 K.

1-7   Use the program `Beattie` to calculate the molar volume of CO_2 at 100 bar and 398.17 K.

1-8   Use the program `Anderko` to calculate the molar volume of CO_2 at 100 bar and 398.7 K.

1-9   Use the program `Keenan1` to calculate the molar volume of steam at 2 bar and 200 °C.

1-10   Use the program `Keenan1` to calculate the molar volume of steam at 500 bar and 1300 °C. Compare with the ideal calculation.

Data Fitting

1-11   Data are given below which express the expansion coefficient α for water as a function of temperature at 1 bar. Use the program `Linreg` to fit these data to the cubic function of temperature: $\alpha/(10^{-5}\ \mathrm{K}^{-1}) = c_0 + c_1 T + c_2 T^2 + c_3 T^3$.

T/K	$\alpha/(\mathrm{K}^{-1}\ 10^{-5})$
298.2	25.53
323.2	46.24
348.2	61.39
373.2	74.86
398.2	88.36
423.2	102.71
448.2	118.70
473.2	137.62

1-12   Data are given below which express the compressibility coefficient β of water as a function of pressure at 423.2 K. Use the program `Linreg` to fit these data to the cubic function of pressure: $\beta/(10^{-6}\ \mathrm{bar}^{-1}) = c_0 + c_1 P + c_2 P^2 + c_3 P^3$.

P/bar	$\beta/(\mathrm{bar}^{-1}\ 10^{-6})$
1	63.09
500	50.41
1000	42.83
2000	34.00
3000	27.50
4000	23.30
5000	20.30

1-13 Use the program `Linreg` to fit the low-temperature ice heat capacity data quoted below to a polynomial temperature function of the kind:

$$C_{Pm}/(\mathrm{J\,K^{-1}\,mol^{-1}}) = c_0 + c_1 T + c_2 T^2 + c_3 T^3 + c_4 T^4 + c_5 T^5.$$

T/K	$C_{Pm}/(\mathrm{J\,K^{-1}\,mol^{-1}})$	T/K	$C_{Pm}/(\mathrm{J\,K^{-1}\,mol^{-1}})$
10	0.27	150	21.95
20	1.98	160	23.14
30	4.10	170	24.37
40	6.12	180	25.61
50	7.90	190	26.84
60	9.61	200	28.12
70	11.26	210	29.49
80	12.82	220	30.82
90	14.38	230	32.11
100	15.83	240	33.41
110	17.22	250	34.71
120	18.49	260	36.03
130	19.71	270	37.36
140	20.82		

2
Chemical Thermodynamics in Theory

The calculations of chemical thermodynamics center on a few *state functions* and on the dependence of the state functions on a few *state variables*. The state functions include *enthalpy* (H), *entropy* (S), and *Gibbs energy* (G), and the state variables are usually pressure (P), absolute temperature (T), and (in solutions of electrolytes) ionic strength (s). If a chemical reaction is involved, the extent-of-reaction variable (ξ) is also important.

Each chemical component in a system is characterized by a molar (or partial molar) enthalpy, entropy, and Gibbs energy. Molar (or partial molar) Gibbs energies are particularly important; for a component A a partial molar Gibbs energy is represented μ_A and also called a *chemical potential*.

The chemical reaction itself is characterized by an enthalpy represented $\Delta_r H$, an entropy $\Delta_r S$, and a Gibbs energy $\Delta_r G$. The reaction enthalpy, $\Delta_r H$, measures the thermal effect produced internally when the reaction proceeds. The reaction entropy, $\Delta_r S$, determines internal disorder changes accompanying the reaction. And the reaction Gibbs energy, $\Delta_r G$, points the spontaneous direction of the reaction: the rules (for a certain pressure and temperature) are that the spontaneous direction is forward if $\Delta_r G < 0$, backward if $\Delta_r G > 0$, and the special condition $\Delta_r G = 0$ defines chemical equilibrium.

Enthalpies and Gibbs energies, like other kinds of energies, must be calculated with respect to reference values. Thermodynamicists use various devices for this, including the concepts of *formation reactions* and *standard states*. Entropies are different; for many (but not all) components, they can be determined as if they were absolute quantities.

In this chapter, some calculational methods are considered that apply this scheme. The programs `Deltag1`, `Deltag2`, and `Deltag3`, introduced in Sec. 2.1, illustrate the utility of heat capacity data for calculating reaction enthalpies, entropies, and Gibbs energies at high temperatures. In Sec. 2.2, methods for calculating enthalpies, entropies, Gibbs energies, compressibility factors, and fugacities for gaseous components at high temperatures and high pressures are introduced and illustrated in the programs `Peng2`, `Peng3`, and `Keenan2`. Four further programs, `Gplot`, `Hplot`, `Splot`, and `Zplot`, display pressure and temperature dependences with surface plots.

The programs Entropy and Debye in Sec. 2.3 show how to calculate "absolute" molar entropies with calorimetric data and Debye's theory. One way to calculate partial molar quantities is introduced in Sec. 2.4 and illustrated with the program Rho. Chemical potential theory is briefly reviewed in Sec. 2.5, and ideal solution behavior is demonstrated by the programs Henry and Raoult. The topic in Sec. 2.7 is nonideal solution behavior of electrolytes and nonelectrolytes, and the illustrative programs are Pitzer and Vanlaar. Finally, in Sec. 2.8, equilibrium theory is outlined and illustrated with the program Gibbs, which shows how the total Gibbs energy G and the reaction Gibbs energy $\Delta_r G$ depend on the extent-of-reaction variable ξ; at equilibrium G has a minimum value and $\Delta_r G = 0$. The program Vanthoff, also introduced in Sec. 2.8, calculates equilibrium constants at high temperatures.

2.1   At High Temperatures

The key to thermodynamic calculations at high temperatures is heat capacity data represented by a convenient temperature function, as in the polynomial equation (1.10) or the Maier–Kelley equation (1.11). The temperature function permits calculation of the reaction heat capacity $\Delta_r C_P$ and that in turn determines the effects of temperature on the reaction enthalpy $\Delta_r H$ and reaction entropy $\Delta_r S$,

$$\left(\frac{\partial(\Delta_r H)}{\partial T}\right)_{P,\xi} = \Delta_r C_P \tag{2.1}$$

$$\left(\frac{\partial(\Delta_r S)}{\partial T}\right)_{P,\xi} = \frac{\Delta_r C_P}{T}. \tag{2.2}$$

If a cubic version of Eq. (1.10) is applicable to all of the components in a reaction, then

$$\Delta_r C_P = \Delta_r c_0 + (\Delta_r c_1)T + (\Delta_r c_2)T^2 + (\Delta_r c_3)T^3,$$

where $\Delta_r c_0$, $\Delta_r c_1$, $\Delta_r c_2$, and $\Delta_r c_3$ are changes in the coefficients c_0, c_1, c_2, and c_3 for the reaction. Substitution from this for $\Delta_r C_P$ in Eqs. (2.1) and (2.2) and integration between a reference temperature T_0 (usually 298.15 K) and T then leads to

$$\Delta_r H(T) = \Delta_r H(T_0) + (\Delta_r c_0)(T - T_0) + (\Delta_r c_1/2)(T^2 - T_0^2)$$
$$+ (\Delta_r c_2/3)(T^3 - T_0^3) + (\Delta_r c_3/4)(T^4 - T_0^4), \tag{2.3}$$

and

$$\Delta_r S(T) = \Delta_r S(T_0) + (\Delta_r c_0)\ln(T/T_0) + \Delta_r c_1(T - T_0)$$
$$+ (\Delta_r c_2/2)(T^2 - T_0^2) + (\Delta_r c_3/3)(T^3 - T_0^3). \tag{2.4}$$

TABLE 2.1. Coefficients in a cubic heat capacity equation with C_{Pm} in $J\,K^{-1}\,mol^{-1}$

Gas	c_0	$c_1/10^{-2}$	$c_2/10^{-5}$	$c_3/10^{-9}$	$T_{max}/K(^*)$
O_2	25.46	1.519	−0.7150	1.311	1800
H_2	29.09	−0.1915	0.4001	−0.8699	1800
N_2	28.88	−0.1570	0.8075	−2.871	1800
Cl_2	28.54	2.389	−2.137	6.473	1500
HCl	30.31	−0.7615	1.326	−4.335	1500
CO_2	22.24	5.977	−3.499	7.464	1800
SO_2	25.76	5.791	−3.809	8.606	1800
CO	28.14	0.1674	0.5368	−2.220	1800
H_2O	32.22	0.1922	1.055	−3.593	1800
NH_3	27.55	2.563	0.9901	−6.686	1500
CH_4	19.87	5.0251	1.268	−11.00	1500
CH_3OH	19.04	9.146	−1.218	−8.033	1000
CCl_4	51.21	14.23	−12.53	36.94	1500

Source: S.I. Sandler, 1989, p. 580.
*Coefficients are valid from 273 K to the listed T_{max}.

From these two results, the reaction Gibbs energy $\Delta_r G$ is calculated with

$$\Delta_r G(T) = \Delta_r H(T) - T\Delta_r S(T). \tag{2.5}$$

We use these equations in the program Deltag1 to calculate standard reaction enthalpies, entropies, and Gibbs energies for a reaction at a high temperature T, given the following data for each component A participating in the formation reaction: the standard enthalpy of formation $\Delta_f H_A^0$ at T_0, the standard entropy S_A^0 at T_0, the standard Gibbs energy of formation, $\Delta_f G_A^0$ at T_0, and the temperature-independent heat capacity parameters c_0, c_1, c_2 and and c_3. These data, values of $\Delta_f H_A^0$, S_A^0, and $\Delta_f G_A^0$ at 298.15 K and 1 bar, and the heat capacity coefficients for selected gases listed in Table 1.1, are all included in the data file Chap2.m (see Table 2.1). The Deltag programs read data from this file.

Example 2-1. Use the program Deltag1 to calculate the standard Gibbs energy of formation for the $H_2O(g)$ formation reaction,

$$H_2(g) + \tfrac{1}{2}O_2(g) \to H_2O(g),$$

at 1000 K.

Answer. Deltag1 requires only that you enter a value for the temperature T on the fourth line of code and a statement of the reaction in reactionList on the sixth line of code,

```
reactionList = H2Og - H2g - .5 O2g;
```

The program calculates $\Delta_f G^0(1000) = -192653 \text{ J mol}^{-1}$. Chase, et. al. (1986), whose calculation is more refined, report $\Delta_f G^0(1000) = -192590 \text{ J mol}^{-1}$.

A more convenient, but less accurate, approach to the same calculation utilizes the Maier–Kelley equation (1.11) to represent the temperature dependence of heat capacities. Equations corresponding to Eqs. (2.3) and (2.4) are then

$$\Delta_r H(T) = \Delta_r H(T_0) + (\Delta_r a)(T - T_0) + (\Delta_r b/2)(T^2 - T_0^2)$$
$$- (\Delta_r c)(1/T - 1/T_0), \tag{2.6}$$

$$\Delta_r S(T) = \Delta_r S(T_0) + (\Delta_r a)\ln(T/T_0) + (\Delta_r b)(T - T_0)$$
$$- (\Delta_r c/2)(1/T^2 - 1/T_0^2), \tag{2.7}$$

and $\Delta_r G(T)$ is again calculated with Eq. (2.5). Values of the heat capacity parameters a, b, and c for some components are listed in Table 2.2, and these data are also included in the data file Chap2.m. This approach is implemented in the program Deltag2.

TABLE 2.2. Coefficients in the Maier–Kelley heat capacity equation for selected components, with C_{Pm} in $\text{J K}^{-1} \text{mol}^{-1}$

Component	a	$b/10^{-3}$	$c/10^5$	$T_{max}/\text{K}(*)$
O_2(g)	29.86	4.184	−1.67	3000
H_2(g)	27.28	3.26	0.50	3000
CH_4(g)	23.64	47.86	−1.92	1500
CO(g)	28.41	4.10	−0.46	3000
CO_2(g)	44.22	8.79	−8.62	2500
H_2O(g)	30.54	10.30	0	2750
C(s) (graphite)	16.86	4.77	−8.54	2500
CaO(s)	49.62	4.52	−6.94	2000
SiO_2(s) (α quartz)	46.94	3.30	−11.3	848
$CaCO_3$(s) (calcite)	104.5	21.9	−25.9	1200
$CaSiO_3$(s)	111.5	15.9	−27.3	1800
$Ca(OH)_2$(s)	105.3	11.94	−19.0	700
$NaHCO_3$(s)	44.89	143.9	0	500
Na_2CO_3(s)	11.02	244.0	−24.5	723

Sources: H.C. Helgeson, et. al., 1978; I. Barin and O. Knacke, 1979.
*Coefficients are valid from 298 K to the T_{max} listed.

Example 2-2. Use the program Deltag2 to calculate $\Delta_r G^0$ for the H_2O(g) formation reaction at 1000 K.

Answer. The program `Deltag2` runs the same way as `Deltag1`. Enter a value for T on the sixth line of code, and specify the reaction in `reaction List` on the fourth line of code. The program calculates $\Delta_r G(1000) = -192744 \, \text{J mol}^{-1}$.

The effects of temperature changes on a reaction Gibbs energy $\Delta_r G$ are determined by

$$\left(\frac{\partial(\Delta_r G)}{\partial T}\right)_{P,\xi} = -\Delta_r S. \tag{2.8}$$

If $\Delta_r S$ expressed by Eq. (2.7) is substituted in this equation, and the resulting equation integrated from T_0 to T, an equation defining the Gibbs energy function $\Delta_r G(T)$,

$$\Delta_r G(T) = A + BT + CT \ln T + DT^2 + E/T, \tag{2.9}$$

is obtained with

$$A = \Delta_r H^0(T_0) - (\Delta_r a)T_0 - (\Delta_r b/2)T_0^2 + \Delta_r c/T_0 \tag{2.10}$$

$$B = \Delta_r a - \Delta_r S^0(T_0) + (\Delta_r a)\ln T_0 + (\Delta_r b)T_0 - (\Delta_r c)/2T_0^2 \tag{2.11}$$

$$C = -\Delta_r a \tag{2.12}$$

$$D = -\Delta_r b/2 \tag{2.13}$$

$$E = -\Delta_r c/2. \tag{2.14}$$

Gibbs energy calculations are handled this way in the program `Deltag3`.

Example 2-3. Use the program `Deltag3` to calculate $\Delta_r G^0$ for the $H_2O(g)$ formation reaction at 1000 K.

Answer. `Deltag3` does the same calculation as `Deltag2` but in a different way. Enter data in `Deltag3` as in `Deltag2` and again calculate $\Delta_r G^0(1000) = -192744 \, \text{J mol}^{-1}$ for the $H_2O(g)$ formation reaction at 1000 K.

2.2   At High Temperatures and High Pressures

The lesson taught in the last section is that the calculations of chemical thermodynamics can be extended to high temperatures by taking advantage of heat capacity data. Here we find that if a good nonideal gas law is available thermodynamic calculations at high pressures are also possible.

Suppose, for example, we need to calculate entropies of gases at high temperatures and also at high pressures. We turn to the entropy temperature and pressure derivatives,

$$\left(\frac{\partial S}{\partial T}\right)_P = \frac{C_P}{T} \tag{2.15}$$

$$\left(\frac{\partial S}{\partial P}\right)_T = -\alpha V, \tag{2.16}$$

where

$$\alpha = \frac{1}{V}\left(\frac{\partial V}{\partial T}\right)_P$$

is the *thermal expansion coefficient*. We can integrate Eq. (2.15) with the heat capacity C_P expressed as a function of temperature of the kind seen in the last section, and integration of Eq. (2.16) is possible with α and V defined by a nonideal gas law.

We will not give the details of such a calculation, but the procedure developed by Sandler (1989) is carried out in the program Peng2. Sandler's calculation is based on the Peng–Robinson nonideal gas law, Eq. (1.6), and it calculates *departure functions* for nonideal gases. For entropy, the "departure" is the difference between the molar entropy $S_m(P, T)$ of the real gas at the pressure P and temperature T and the molar entropy $S_{mig}(1, 298.2)$ of the same gas behaving ideally at 1 bar and 298.2 K ("ig" stands for ideal gas). Peng2 also calculates the enthalpy departures defined similarly, beginning with the enthalpy temperature and pressure derivatives,

$$\left(\frac{\partial H}{\partial T}\right)_P = C_P \tag{2.17}$$

$$\left(\frac{\partial H}{\partial P}\right)_T = V(1 - \alpha T). \tag{2.18}$$

Data required by Peng2 are values of the critical constants P_c and T_c for the gas, the acentric factor ω (Table 1.2), and the parameters c_0, c_1, c_2 and c_3 that fit heat capacity data to a cubic temperature function (Table 2.1). The next example demonstrates.

Example 2-4. Use the program Peng2 to calculate entropy departures for CO_2 at 400 K, first for 50 bar and then for 100 bar. The difference between these two results is the entropy change in CO_2 when it is compressed at 400 K from 50 to 100 bar. Compare this entropy change with that calculated for an ideal gas.

Answer. Enter 50 bar and 400 K for the pressure and temperature on the third and fourth lines of code in `Peng2`, and identify the gaseous component five lines later with

$$gas = CO2g;$$

Calculate

$$S_m(50,400) - S_{mig}(1,298.2) = -23.48 \text{ J K}^{-1} \text{ mol}^{-1}$$

for the entropy departure. Then enter 100 bar for the pressure and calculate

$$S_m(100,400) - S_{mig}(1,298.2) = -31.89 \text{ J K}^{-1} \text{ mol}^{-1}.$$

Thus

$$\Delta S_m = S_m(100,400) - S_m(50,400) = -8.41 \text{ J K}^{-1} \text{ mol}^{-1}.$$

The same calculation for an ideal gas is

$$\Delta S = R \ln \frac{100 \text{ bar}}{50 \text{ bar}} = -5.8 \text{ J K}^{-1} \text{ mol}^{-1}.$$

The ideal calculation is considerably in error.

In Sec. 1.4 we used the Keenan–Keyes–Hill–Moore nonideal gas law to calculate molar volumes and densities of steam at high pressures and high temperatures. We follow this analysis further now and calculate enthalpies and entropies of steam.

Keenan, Keyes, Hill, and Moore organized their further calculations around the Helmholtz energy A expressed as a function of the density ρ. Since ρ depends on the volume V and temperature T, we can also treat A as the function $A(V,T)$. Thermodynamics then provides the equations needed to calculate the entropy S, internal energy U and enthalpy H,

$$S = -\left(\frac{\partial A}{\partial T}\right)_V \tag{2.19}$$

$$U = A + TS$$

$$= A - T\left(\frac{\partial A}{\partial T}\right)_V \tag{2.20}$$

$$H = U + PV$$

$$= A - T\left(\frac{\partial A}{\partial T}\right)_V - V\left(\frac{\partial A}{\partial V}\right)_T. \tag{2.21}$$

In the last line $P = -(\partial A/\partial V)_T$ is used to calculate the pressure P. The program `Keenan2`, demonstrated in the next example, carries out the Keenan–Keyes–Hill–Moore internal energy, enthalpy and entropy calculations.

Example 2-5. Use the program `Keenan2` to calculate the entropy and enthalpy of steam at 500.0 °C and 200 bar.

Answer. Enter 200 bar and 773.2 K for the pressure and temperature on lines three and four of code in `Keenan2`, and calculate the following values for the specific internal energy u, enthalpy h and entropy s ("specific" means per unit mass, in this case, per gram),

$$u = 2942.8 \ \mathrm{J\,g^{-1}}$$

$$h = 3238.3 \ \mathrm{J\,g^{-1}}$$

$$s = 6.140 \ \mathrm{J\,K^{-1}\,g^{-1}}.$$

These results are calculated with respect to zero values at the water triple point.

The molar Gibbs energy of an ideal gas at the pressure P and temperature T is calculated with

$$G_m = G_m^0 + RT \ln (P/P^0),$$

in which P^0 is a reference pressure which we assume to be equal to (exactly) 1 bar, and G_m^0 is therefore the molar Gibbs energy at that pressure. If the gas is nonideal, an equation of the same form applies, but the pressure P is replaced with the *fugacity* f,

$$G_m = G_m^0 + RT \ln (f/P^0). \tag{2.22}$$

A factor γ called a *fugacity coefficient* connects a fugacity f and the corresponding pressure P,

$$f = \gamma P. \tag{2.23}$$

A further adaptation of the Peng programs called `Peng3` calculates fugacities and fugacity coefficients.

Example 2-6. Use the program `Peng3` to calculate the fugacity and fugacity coefficient of steam at 500.0 °C and 200 bar. The measured value of the fugacity coefficient of steam under these conditions is 0.85.

Answer. `Peng3` requires the same input as `Peng1`. Enter the required data on the third and fourth lines of code, and run the program to obtain $\gamma = 0.83$ for the fugacity coefficient and $f = 166.0$ bar for the fugacity.

The program `Peng2` is easily adapted so it plots three-dimensional surfaces representing the dependence of the thermodynamic functions G, H, S, and Z on the thermodynamic variables P and T for nonideal gases. This is done in the series of programs `Gplot`, `Hplot`, `Splot`, and `Zplot`. Run these programs and note particularly that with increasing pressure G, H, and S increase, decrease (slightly) and decrease, and with increasing temperature the same functions decrease, increase and increase.

2.3   Calorimetric Entropies

We saw in the last section that heat capacity data permit the extension of thermodynamic calculations to high temperatures. We can also use heat capacity data for low temperature calculations. The most important application of this kind determines calorimetric molar entropies according to the Third Law prescription

$$S_m(T) = \int_0^T \frac{C_{Pm}(T')}{T'} \, dT'. \tag{2.24}$$

The integral is partly evaluated with a series of heat capacity measurements beginning at about 10 K and extending to a higher temperature T, usually around room temperature.

Heat capacity measurements are difficult at very low temperatures, but that temperature range need not be covered by measurements because a theory of the solid state due to Debye shows how to extrapolate heat capacity calculations to absolute zero. Thus the full calculation of a calorimetric entropy is

$$S_m(T) = \int_0^{T_0} \frac{C_{Pm}(T')}{T'} \, dT' + \int_{T_0}^T \frac{C_{Pm}(T')}{T'} \, dT', \tag{2.25}$$

where the first integral, with T_0 typically equal to about 10 K, is calculated with the Debye extrapolation.

The program `Entropy` calculates the second integral in Eq. (2.25), with heat capacities expressed as polynomial temperature functions as in Eq. (1.10). The equation for the second integral is

$$\int_{T_0}^T \frac{C_{Pm}(T')}{T'} \, dT' = [c_0 \ln T + c_1 T + (c_2/2)T^2 + \cdots + (c_n/n)T^n]$$

$$- [c_0 \ln T_0 + c_1 T_0 + (c_2/2)T_0^2 + \cdots + (c_n/n)T_0^n]. \tag{2.26}$$

The Debye extrapolation can often be made by simply assuming that heat capacities of nonmetallic, nonmagnetic solids at low temperatures are proportional to T^3, that is,

$$C_{Pm}(T) = DT^3,$$

with D a constant that depends on the identity of the solid but not on temperature. (For more on how to calculate this constant, see the program Debye.) Thus the first integral in Eq. (2.25) becomes

$$\int_0^{T_0} \frac{C_{Pm}(T')\,dT'}{T'} = \frac{DT_0^3}{3}$$

$$= \frac{C_{Pm}(T_0)}{3}, \qquad (2.27)$$

and the full calorimetric entropy calculation is

$$S_m(T) = \frac{C_{Pm}(T_0)}{3} + \int_{T_0}^{T} \frac{C_{Pm}(T')}{T'}\,dT'. \qquad (2.28)$$

The program Debye provides a more accurate calculation of the Debye extrapolation.

Example 2-7. Heat capacity data taken by Hemingway, et. al. (1977) on the mineral magnesite ($MgCO_3$) fit the polynomial temperature function,

$$C_{Pm}(T)/(\text{J K}^{-1}\,\text{mol}^{-1}) = 3.811 - 3.962 \times 10^{-1}\,T + 1.132 \times 10^{-2}\,T^2$$

$$- 6.975 \times 10^{-5}\,T^3 + 1.853 \times 10^{-7}\,T^4$$

$$- 1.830 \times 10^{-10}\,T^5,$$

from $T = 12.73$ to 332.19 K at 1 bar. The heat capacity of magnesite at 12.73 K is 0.023 J K^{-1} mol^{-1}. Use the program Entropy to calculate the calorimetric entropy of magnesite at 298.15 K and 1 bar.

Answer. Enter data in Entropy as instructed in the program and calculate

$$\int_{12.73}^{298.15} \frac{C_{Pm}(T')}{T'}\,dT' = 64.83 \text{ J K}^{-1}\,\text{mol}^{-1},$$

and then according to Eq. (2.28),

$$S_m(298.15) = 0.023/3 + 64.83$$

$$= 64.84 \text{ J K}^{-1}\,\text{mol}^{-1}.$$

2.4 Partial Molar Quantities

If you mix n_A moles of component A with n_B moles of B to form $n_A + n_B$ moles of a solution, the total volume V of the solution is

$$V = n_A V_A + n_B V_B,$$

in which V_A and V_B are *partial molar volumes* defined

$$V_A = \left(\frac{\partial V}{\partial n_A}\right)_{P,T,n_B} \tag{2.29}$$

$$V_B = \left(\frac{\partial V}{\partial n_B}\right)_{P,T,n_A}, \tag{2.30}$$

and the molar volume is

$$V_m = x_A V_A + x_B V_B, \tag{2.31}$$

in which x_A and x_B are mole fractions.

This same approach can be applied to calculation of other thermodynamic state functions. For example, the molar Gibbs energy G_m of a solution containing components A and B is

$$G_m = x_A \mu_A + x_B \mu_B,$$

where μ_A and μ_B are *chemical potentials*, defined

$$\mu_A = \left(\frac{\partial G}{\partial n_A}\right)_{P,T,n_B} \tag{2.32}$$

$$\mu_B = \left(\frac{\partial G}{\partial n_B}\right)_{P,T,n_A}. \tag{2.33}$$

Partial molar quantities are often determined by first calculating an *apparent molar quantity*. An example is the *apparent molar volume*,

$$V_\phi = \frac{V - n_B V_B^*}{n_A},$$

with $V = n_A V_A + n_B V_B$ the total volume and V_B^* the molar volume of pure B. If A and B play the roles of solute and solvent, and $cm^3 \, mol^{-1}$ units are used for V_A, V_B and V_ϕ, then

$$V_\phi = \frac{M_A}{\rho} + \frac{(1000 \text{ g kg}^{-1})}{m_A}\left(\frac{1}{\rho} - \frac{1}{\rho_B^*}\right), \tag{2.34}$$

where ρ and ρ_B^* represent densities (in $g\,cm^{-3}$) of the solution and the pure solvent, and M_A and m_A the molar mass (in $g\,mol^{-1}$) and molality (in $mol\,kg^{-1}$) for A. From V_ϕ, the two partial molar volumes V_A and V_B are calculated with respect to V_A^* and V_B^* for the pure components according to

$$V_A = V_\phi + m_A\left(\frac{\partial V_\phi}{\partial m_A}\right)_{P,T,n_B} \tag{2.35}$$

$$V_B = \frac{M_B m_A^2}{(1000 \text{ g kg}^{-1})}\left(\frac{\partial V_\phi}{\partial m_A}\right)_{P,T,n_B}. \tag{2.36}$$

The program Rho does this calculation for ethanol-water solutions beginning with accurate density data. Run the program and note the complicated dependence of V_A and V_B on the composition of the solution.

2.5   Chemical Potential Theory

Chemical potentials are important in chemical thermodynamics because they permit calculation of reaction Gibbs energies. For the generic reaction

$$aA + bB \rightarrow rR + sS,$$

the Gibbs energy is

$$\Delta_r G = r\mu_R + s\mu_S - a\mu_A - b\mu_B, \tag{2.37}$$

and $\Delta_r G$ points the spontaneous direction of the reaction, as mentioned at the beginning of the chapter.

Chemical potentials are defined formally as derivatives [Eqs. (2.32) and (2.33)] but are usually calculated from *activities*, the general connection between the chemical potential μ_A and the activity a_A for a component A being

$$\mu_A = \mu_A^0 + RT \ln a_A, \tag{2.38}$$

in which μ_A^0, a *standard chemical potential*, evaluates μ_A, when $a_A = 1$ and is independent of composition, but dependent on pressure and temperature.

Equation (2.38) expresses all of the composition dependence of the chemical potential μ_A in the activity a_A. If A is a solute, composition is further expressed by relating the activity a_A to the concentration of A. One way to do this is with the equation

$$a_A = \gamma_A m_A / m^0, \tag{2.39}$$

where m_A is the molality of A (moles of A per kilogram of solvent), γ_A is an *activity coefficient* which depends on m_A and perhaps on concentrations of other components in the solution, and m^0 is a reference molality, taken to be $1 \, \text{mol kg}^{-1}$. Two other activity equations commonly used for solutes are

$$a_A = \gamma_A C_A / C^0, \tag{2.40}$$

and

$$a_A = \gamma_A x_A, \tag{2.41}$$

with C_A and x_A representing concentration as a molarity (moles of A per liter of solution) and a mole fraction (moles of A per mole of solution), $C^0 = 1 \, \text{mol L}^{-1}$ and the γ's are two more activity coefficients. Equations (2.39) and (2.40) are not suitable for solvent components, but Eq. (2.41) is.

If A is a component of a gas-phase mixture

$$a_A = \gamma_A P_A / P^0, \tag{2.42}$$

where P^0 is a reference pressure equal to 1 bar and γ_A is another activity coefficients. [You may have to remind yourself that the γ_A's in Eqs. (2.39)–(2.42) are all different quantities; later we will recognize the differences explicitly.]

Combination of Eqs. (2.37) and (2.38) leads to an important equation relating the reaction Gibbs energy $\Delta_r G$ to the activities of the components participating in the reaction,

$$\Delta_r G = \Delta_r G^0 + RT \ln Q, \qquad (2.43)$$

in which Q is the *activity quotient*,

$$Q = \frac{a_R^r a_S^s}{a_A^a a_B^b},$$

for the reaction and $\Delta_r G^0$ is the reaction's *standard Gibbs energy*, the Gibbs energy for the reaction when all of the activities have unit values.

Chemical equilibrium is defined by $\Delta_r G = 0$ and by an equilibrium value of the activity quotient, represented by K and called an *equilibrium constant*. Constraining Eq. (2.43) with equilibrium conditions we see that

$$\Delta_r G^0 = -RT \ln K. \qquad (2.44)$$

2.6   Ideal Solutions

In ideal solutions, relevant activity coefficients have values of one. Several experimental criteria are also important for identifying ideal solution behavior. One, called *Raoult's law*, and usually applied to an ideal solvent component, is

$$p_A = p_A^* x_A, \qquad (2.45)$$

in which p_A is the equilibrium vapor pressure of A over the solution and p_A^* is the vapor pressure of pure A. Ideal solute components obey *Henry's law*, either

$$p_A = k_A x_A, \qquad (2.46)$$

or

$$p_A = k_A' m_A, \qquad (2.47)$$

depending on the concentration scale preferred (mole fraction x_A or molality m_A). In these statements k_A and k_A' are two different *Henry's law constants*, related by

$$k_A/k_A' = (1000 \text{ g kg}^{-1})/M_B, \qquad (2.48)$$

in which M_B is the molar mass of the solvent expressed in the units $g\,mol^{-1}$.

If Raoult's law, Eq. (2.45), is obeyed, the corresponding chemical potential calculation, usually for a solvent components, is

$$\mu_A = \mu_{Aig}^0 + RT \ln(p_A^*/p^0) + RT \ln x_A, \qquad (2.49)$$

with μ_{Aig}^0 the standard chemical potential for A behaving as an ideal gas and p^0 a reference pressure, assumed as before to be equal to 1 bar. We do not attempt to calculate μ_{Aig}^0, but use it as an energy reference for chemical potential calculations.

Chemical potentials for ideal solute components are calculated with

$$\mu_A = \mu_{Aig}^0 + RT \ln(k_A/p^0) + RT \ln x_A, \qquad (2.50)$$

if Henry's law, Eq. (2.46), is obeyed and with

$$\mu_A = \mu_{Aig}^0 + RT \ln(k_A'm^0/p^0) + RT \ln(m_A/m^0) \qquad (2.51)$$

if the other Henry's law, Eq. (2.47), is followed. In the second equation m^0 is a reference molality, assumed as before to be equal to $1\,mol\,kg^{-1}$.

The program Raoult displays plots of Eqs. (2.49) and (2.50) for solutions of acetone and carbon disulfide. The plot shows Henry's law and Raoult's law behavior (as seen in chemical potential calculations) and also shows how real behavior passes between the two ideal modes. Data used to make the calculation, and similar data for other systems, are listed in Tables 2.3 and 2.4. Van Laar parameters, listed in Table 2.4, are discussed in the next section.

The program Henry plots the ideal equation (2.51) and compares it with real behavior. As in the plot made with Raoult the solute approaches ideal behavior at low concentrations.

TABLE 2.3. Vapor pressures and molar masses for some volatile components

Component	p_A^*/Torr	$M_A/(g\,mol^{-1})$	T/K
Acetone	231.6	58.08	298.2
Acetone	615.9	58.08	323.2
Acetone	868.7	58.08	333.2
Benzene	391.4	78.114	333.2
Carbon disulfide	858.5	76.131	323.2
Ethanol	59.3	46.069	298.2
Methanol	127.6	32.042	298.2
Water	23.8	18.015	298.2

Source: Reid, Prausnitz, and Poling, 1987, Appendix A.

TABLE 2.4. Van Laar parameters and Henry's law constants for some binary solutions

A	B	a	b	k_A/bar	k_B/bar	T/K
Acetone	Carbon disulfide	1.79	1.28	4.91	4.12	323.2
Acetone	Water	0.405	0.405	1.74	0.782	333.2
Acetone	Water	1.89	1.66	2.04	0.167	298.2
Ethanol	Water	1.54	0.97	0.369	0.0837	298.2
Methanol	Water	0.58	0.46	0.304	0.0503	298.2

Sources: Sandler, 1989, p. 328, and Henry's law constants calculated with Eq. (2.60).

2.7   Nonideal Solutions

Nonideal solutions of all kinds are characterized by activity coefficients that are not equal to one. The necessary activity coefficients are added to Eqs. (2.49) to (2.51) as multplied factors preceding the concentration factors in the logarithm terms. To accomodate the different forms of the equations three different activity coefficients are needed. Thus the ideal equation (2.49) becomes the nonideal equation

$$\mu_A = \mu_{Aig}^0 + RT\ln(p_A^*/p^0) + RT\ln[\gamma_A(r)x_A]. \tag{2.52}$$

[This equation has its origins in Raoult's law; hence the "r" in $\gamma_A(r)$.] Ideal equations (2.50) and (2.51) become the nonideal equations

$$\mu_A = \mu_{Aig}^0 + RT\ln(k_A/p^0) + RT\ln[\gamma_A(h1)x_A] \tag{2.53}$$

$$\mu_A = \mu_{Aig}^0 + RT\ln(k_A'm^0/p^0) + RT\ln[\gamma_A(h2)m_A/m^0]. \tag{2.54}$$

[These equations originate in the two versions of Henry's law; hence the "h1" and "h2" labels in $\gamma_A(h1)$ and $\gamma_A(h2)$.]

Many methods for measuring and calculating activity coefficients have been developed. We mention only a few. When binary mixtures of volatile components are involved the approximate equations

$$\gamma_A = p_A/x_A p_A^* \quad \text{and} \quad \gamma_B = p_B/x_B p_B^* \tag{2.55}$$

are useful. These equations are used in the programs Henry and Raoult. They are valid if the components behave ideally in the vapor phase, as is usually the case to a good approximation.

A set of empirical equations introduced by van Laar is often used to fit activity coefficient data for binary mixtures,

$$\ln\gamma_A = \frac{ax_B^2}{(x_B + ax_A/b)^2} \tag{2.56}$$

$$\ln\gamma_B = \frac{bx_A^2}{(x_A + bx_B/a)^2}, \tag{2.57}$$

in which γ_A and γ_B are determined by Eq. (2.55) and a and b are *van Laar parameters*. If data for γ_A and γ_B are available, values of a and b can be calculated with

$$a = \left(1 + \frac{x_B \ln \gamma_B}{x_A \ln \gamma_A}\right)^2 \ln \gamma_A \qquad (2.58)$$

$$b = \left(1 + \frac{x_A \ln \gamma_A}{x_B \ln \gamma_B}\right)^2 \ln \gamma_B. \qquad (2.59)$$

Van Laar parameters are useful for representing nonideality with the activity coefficient equations (2.56) and (2.57). They also provide a simple route to calculation of Henry's law constants. For a component A,

$$k_A = p_A^* e^a. \qquad (2.60)$$

The program Vanlaar calculates the van Laar parameters a and b according to Eqs. (2.58) and (2.59).

Examples of nonideal solutions mentioned so far have concerned non-electrolytes. In a formal sense the chemical thermodynamics of electrolytes is similar. The general activity equation (2.38) again applies and activities of solutes are expressed with Eq. (2.39) or (2.40).

Electrolytes in solutions form ions, however, which cannot be separated from each other in thermodynamic measurements and calculations, except as an approximation. For this reason, activity coefficients are often calculated as geometric means of contributions made by anion and cation components. The electrolyte NaCl, for example, forms the ions Na^+ and Cl^- and the NaCl mean activity coefficient is defined $\gamma_\pm = (\gamma_{Na^+}\gamma_{Cl^-})^{1/2}$. Mean activity coefficients for $CaCl_2$, Na_2SO_4 and $CaSO_4$ and $\gamma_\pm = (\gamma_{Ca^{2+}}\gamma_{Cl^-}^2)^{1/3}$, $(\gamma_{Na^+}^2\gamma_{SO_4^{2-}})^{1/3}$ and $(\gamma_{Ca^{2+}}\gamma_{SO_4^{2-}})^{1/2}$.

Activity coefficients for dilute electrolyte solutions are calculated with a theory introduced by Debye and Hückel. The Debye–Hückel result for *very* dilute electrolyte solutions is

$$\ln \gamma_\pm = -|z_+|\,|z_-|A\sqrt{s}, \qquad (2.61)$$

in which z_+ and z_- are cation and anion charge numbers, A is a constant evaluated by the theory and s is the ionic strength defined

$$s = \frac{1}{2}\sum_i m_i z_i^2,$$

with the m_i's and z_i's representing molalities and charge numbers for *all* ionic components in the solution. The constant A depends on the identity of the solvent and slightly on the temperature: its value for aqueous solutions at $25\,°C$ is $1.171\ kg^{1/2}\,mol^{-1/2}$.

An extension of the Debye–Hückel theory to dilute, but not necessarily

TABLE 2.5. Debye–Hückel size parameters for single ions

Ions	r_B/nm
K^+, Cl^-, I^-	
OH^-, $H_2Citrate^-$	
Na^+, SO_4^{2-}, PO_4^{3-}, HPO_4^{2-}	0.40
CO_3^{2-}, HCO_3^-, Acetate$^-$,	
$HCitrate^{2-}$, $H_2PO_4^-$	0.45
Ba^{2+}, Citrate^{3-}	0.50
Ca^{2+}, Zn^{2+}, Benzoate$^-$	0.60
Mg^{2+}	0.80
H^+, Al^{3+}	0.90

Source: J. Kielland, 1937.

very dilute, solutions leads to the equation

$$\ln \gamma_\pm = -\frac{|z_+| \, |z_-| A\sqrt{s}}{1 + B\sqrt{s}},$$

(2.62)

where B is an empirical parameter whose value depends on the identity of the electrolyte. This equation has a useful approximate form which calculates the activity coefficient for an individual ion. For an ion B with charge number z_B, the approximate equation is

$$\ln \gamma_B = -\frac{Az_B^2 \sqrt{s}}{1 + r_B b\sqrt{s}},$$

(2.63)

in which b is another constant evaluated by the theory ($= 3.282 \text{ kg}^{1/2}$ $\text{mol}^{-1/2} \text{nm}^{-1}$ for aqueous solutions at 25 °C) and r_B, called the *size parameter*, cannot be determined independently, so it is treated as an empirical parameter. Values of r_B for a selection of ions are listed in Table 2.5.

Equation (2.61) is valid at ionic strengths less than about 0.01 mol kg^{-1}, and Eqs. (2.62) and (2.63) are useful at ionic strengths below about 0.1 mol kg^{-1}. Pitzer and his colleagues have developed a semiempirical method for calculating mean activity coefficients of electrolytes at high ionic strengths, to 1 mol kg^{-1} and beyond. Pitzer's equation takes the form of a virial expansion resembling the virial nonideal gas law, Eq. (1.1). In the latter the variable of the expansion is $1/V_m$; in Pitzer's equation the expansion variable is m, the total molal concentration of the electrolyte. The form of the equation is simply

$$\ln \gamma_\pm = f_0 + f_1 m = f_2 m^2,$$

(2.64)

in which f_0, f_1, and f_2—the first, second, and third virial coefficients—are empirical parameters that vary from one electrolyte to another.

The program $\texttt{Pitzer}$ implements this procedure. To run the program

TABLE 2.6. Pitzer's parameters for some 1-1, 2-1, 1-2, and 2-2 electrolytes at 25 °C

Electrolyte	$\beta^{(0)}$	$\beta^{(1)}$	$\beta^{(2)}$	$C_{MX}{}^\phi$	$m_{max}/(\text{mol kg}^{-1})$
NaCl	0.0765	0.2664	0	0.00127	6
NaNO$_3$	0.0641	0.1015	0	−0.0049	5
KOH	0.1298	0.3200	0	0.0041	5.5
AgNO$_3$	−0.0856	0.0025	0	0.00591	6
MgCl$_2$	0.3523	1.6815	0	0.00519	4.5
CaCl$_2$	0.3159	1.6140	0	−0.0034	2.5
BaCl$_2$	0.2628	1.4963	0	−0.01938	1.8
Na$_2$SO$_4$	0.0196	1.1130	0	0.00570	4.0
Na$_2$CO$_3$	0.1897	0.8460	0	−0.04803	1.5
MgSO$_4$	0.2210	3.343	−37.23	0.0250	3.0
ZnSO$_4$	0.1949	2.883	−32.81	0.0290	3.5
NiSO$_4$	0.1702	2.907	−40.06	0.0366	2.5
CdSO$_4$	0.2053	2.617	−48.07	0.0114	3.5

Source: K.S. Pitzer, *Thermodynamics*, 1995, Appendix 8.

you need only identify on the fourth line of code the electrolyte component for which you want to make the calculation, and then later a molality m if you want to calculate an activity coefficient at a particular concentration, and yMax = 0.1 if a 2-2 electrolyte is involved. A short list of Pitzer's parameters is given in Table 2.6 and the same data are included in the file Chap2.m. A much longer list is available in Pitzer's thermodynamics text (Appendix 8).

2.8 Chemical Equilibrium Theory

As mentioned at the beginning of the chapter, a chemical reaction reaches equilibrium when its Gibbs energy vanishes,

$$\Delta_r G = 0. \tag{2.65}$$

Since $\Delta_r G$ is defined as a derivative with respect to the extent-of-reaction variable ξ,

$$\Delta_r G = \left(\frac{\partial G}{\partial \xi}\right)_{P,T}, \tag{2.66}$$

Eq. (2.65) also tells us that the derivative $(\partial G/\partial \xi)_{P,T}$ has a zero value at equilibrium. That condition occurs when the system's Gibbs energy G reaches a minimum value.

These conclusions are demonstrated by the program Gibbs, which plots both G and its derivative (i.e., $\Delta_r G$) as functions of ξ for a system in which a reaction of the following kind occurs,

$$A(g) \rightarrow R(g) + S(g),$$

involving ideal gas components. The reaction begins with n_0 moles of A and no R or S. When the extent of reaction is ξ molar amounts of A, R, and S are $n_A = n_0 - \xi, n_R = n_S = \xi$, the total molar amount is $n = n_A + n_R + n_S = n_0 + \xi$, and

$$p_A = \frac{n_0 - \xi}{n_0 + \xi} P$$

$$p_R = p_S = \frac{\xi}{n_0 + \xi} P,$$

with P the total pressure. Chemical potentials for the components A, R and S are

$$\mu_A = \mu_A^0 + RT \ln(p_A/p^0)$$

$$\mu_R = \mu_R^0 + RT \ln(p_R/p^0)$$

$$\mu_S = \mu_S^0 + RT \ln(p_S/p^0),$$

and Gibbs energy equations for the reaction are

$$G = n_A \mu_A + n_R \mu_R + n_S \mu_S$$

and

$$\Delta_r G = \mu_R + \mu_S - \mu_A.$$

Evaluating the chemical potentials in terms of ξ and rearranging, these equations become

$$\frac{G(\xi) - G(0)}{RT} = -\xi \ln K + \xi \ln(P/P^0) + (n_0 - \xi) \ln(n_0 - \xi)$$

$$- (n_0 + \xi) \ln(n_0 + \xi) + 2\xi \ln \xi, \tag{2.67}$$

and

$$\frac{\Delta_r G}{RT} = -\ln K + \ln(P/P^0) - \ln(n_0 - \xi) - \ln(n_0 + \xi) + 2 \ln \xi, \tag{2.68}$$

in which $G(0)$ is the initial Gibbs energy of the system (i.e., for n_0 moles of A) and P^0 is the reference pressure 1 bar. These are the equations plotted by the program Gibbs. You can demonstrate Eq. (2.66) directly by differentiating Eq. (2.67) with respect to ξ and comparing it with Eq. (2.68).

Equilibrium constant data are a crucial part of most chemical equilibrium calculations. To obtain equilibrium constants at high temperatures use the *van't Hoff equation*,

$$\left(\frac{\partial \ln K}{\partial T}\right)_P = \frac{\Delta_r H^0}{RT^2}, \tag{2.69}$$

or, alternatively, Eq. (2.44),

$$\Delta_r G^0 = -RT \ln K.$$

It is particulalry convenient to combine the latter equation with $\Delta_r G^0$ expressed as a function of T as in Eq. (2.9). This is done in the program Vanthoff, which plots $\ln K$ vs $1/T$ for reactions of interest.

The van't Hoff equation (2.69) tells us that for endothermic reactions ($\Delta_r H^0 > 0$) equilibrium constants increase with temperature and for exothermic reaction ($\Delta_r H^0 < 0$) decrease with temperature. Thus endothermic reactions tend to dominate at high temperatures and exothermic reactions at low temperatures. The plots made by the program Vanthoff illustrate these trends. (Note that the temperature *reciprocal* $1/T$ is plotted by Vanthoff, so T increases from right to left in the plots.)

It is often legitimate to simplify equilibrium calculations by assuming that $\Delta_r H^0$ in the van't Hoff equation (2.69) is independent of temperature. Then integration of Eq. (2.69) leads to

$$\ln K = \frac{\Delta_r S^0}{R} - \frac{\Delta_r H^0}{RT}. \tag{2.70}$$

This suggests measuring K for a reaction over a range of temperatures and then plotting $\ln K$ vs $1/T$ to obtain $\Delta_r S^0$ and $\Delta_r H^0$ for the reaction. The program Vanthoff does its plotting this way and in the high temperature range covered (500 to 2000 K) the curves are nearly straight lines with intercepts and slopes equal to $\Delta_r S^0/R$ and $-\Delta_r H^0/R$.

2.9   Exercises

At High Temperatures

2-1   Calculate the standard Gibbs energy of formation of $NH_3(g)$ at 1000 K using the program Deltag1.

2-2   Calculate the standard Gibbs energy of formation of $CH_4(g)$ at 1000 K using the program Deltag3.

2-3   Use the program Deltag2 to calculate $\Delta_r H^0$ and $\Delta_r S^0$ for the reaction

$$CaO(g) + H_2O(g) \rightarrow Ca(OH)_2(s)$$

at 350 K.

2-4   Use the program Deltag2 to calculate $\Delta_r H^0$ and $\Delta_r S^0$ for the reaction

$$2\,NaHCO_3(s) \rightarrow Na_2CO_3(s) + H_2O(g) + CO_2(g)$$

at 373.2 K.

2-5   Use the program `Deltag1` to calculate $\Delta_r H^0$, $\Delta_r S^0$, and $\Delta_r G^0$ for the reaction

$$2\,CO(g) + O_2(g) \rightarrow 2\,CO_2(g)$$

at 500 K.

2-6   Use the program `Deltag1` to calculate $\Delta_r H^0$, $\Delta_r S^0$, and $\Delta_r G^0$ for the reaction

$$CH_4(g) + 2\,O_2(g) \rightarrow CO_2(g) + 2\,H_2O(g)$$

at 900 K.

At High Temperatures and High Pressures

2-7   Use the program `Peng2` to calculate the entropy and enthalpy changes in a mole of $O_2(g)$ when it is compressed from 1.00 bar to 100 bar. The temperature increases in the process from $25.00\,^\circ C$ to $150.00\,^\circ C$.

2-8   Use the program `Peng3` to calculate the fugacity and fugacity coefficient of $CH_4(g)$ at $100.0\,^\circ C$ and 100 bar.

2-9   Calculate the Gibbs energy change for a process that compresses $NH_3(g)$ at $100.0\,^\circ C$ from 1.00 bar to 50 bar. Use the program `Peng3` to calculate fugacities.

Calorimetric Entropies

2-10   Heat capacity data taken by Giauque and Powell on solid chlorine between $T_0 = 15.00$ K and $T_m = 172.12$ K (the melting point) at 1 atm fit the cubic polynomial temperature function (see Exercise 2-11),

$$C_{P,Cl_2}(s,\,T) = -12.60 + 1.231\,T - 9.420 \times 10^{-3}\,T^2 + 2.668 \times 10^{-5}\,T^3.$$

Their heat capacity data for the reversible heating of liquid chlorine between $T_m = 172.12$ K and $T_b = 239.05$ K (the boiling point) fit the cubic temperature function

$$C_{P,Cl_2}(l,\,T) = 48.91 + 0.2632\,T - 1.197 \times 10^{-3}\,T^2 + 1.631 \times 10^{-6}\,T^3.$$

The heat capacity of solid chlorine at 15.00 K is $3.72\ J\,K^{-1}\,mol^{-1}$. Giauque and Powell also measured $6406\ J\,mol^{-1}$ for the enthalpy of melting of solid chlorine at $T_m = 172.12$ K and $20410\ J\,mol^{-1}$ for the enthalpy of boiling at $T_b = 239.05$ K and 1 atm. Calculate the calorimetric entropy of chlorine gas at 239.05 K and a atm. Giauque and Powell reported $215.22\ J\,K^{-1}\,mol^{-1}$.

2-11   Heat capacity data for solid and liquid chlorine listed in Table 2.7 were obtained by Giauque and Powell. Use the data and the program `Linreg` to derive cubic temperature functions that fit the data.

TABLE 2.7. Heat capacity data for chlorine

	T/K	$C_{Pm}/(\text{J K}^{-1}\,\text{mol}^{-1})$	T/K	$C_{Pm}/(\text{J K}^{-1}\,\text{mol}^{-1})$
Solid	15	3.724	80	38.618
	20	7.740	90	40.627
	25	12.092	100	42.258
	30	16.694	110	43.806
	35	20.794	120	45.480
	40	23.974	130	47.237
	45	26.736	140	49.078
	50	29.246	150	51.045
	60	33.472	160	53.053
	70	36.317	170	55.103
			172.12	55.522
			(= melting point)	
Liquid	172.12	67.070	210	66.484
	180	67.028	220	66.275
	190	66.902	230	65.982
	200	66.735	240	65.689

2-12 Heat capacity data obtained by Giauque and Stephenson for solid and liquid SO_2 are given in Table 2.8. Giauque and Stephenson also measured 7402 J mol^{-1} and 24940 J mol^{-1} for the molar enthalpy of melting and molar enthalpy of boiling of SO_2 at 1 atm. Use the program Linreg to fit the two sets of heat capacity data to cubic temperature functions. Then use

TABLE 2.8. Heat capacity data for sulfur dioxide

	T/K	$C_{Pm}/(\text{J K}^{-1}\,\text{mol}^{-1})$	T/K	$C_{Pm}/(\text{J K}^{-1}\,\text{mol}^{-1})$
Solid	15	3.473	90	45.731
	20	6.945	100	48.074
	25	11.464	110	50.082
	30	15.857	120	51.882
	35	20.292	130	53.681
	40	24.184	140	55.689
	45	27.656	150	57.823
	50	30.794	160	59.957
	55	33.556	170	62.132
	60	36.066	180	64.517
	70	40.041	190	67.028
	80	43.179	197.64	69.496
			(= melting point)	
Liquid	197.64	87.697	240	86.860
	200	87.738	250	86.651
	210	87.487	260	86.441
	220	87.278	263.08	86.358
	230	87.069	(= boiling point)	

the program Entropy and other data supplied to calculate the entropy of:
(a) solid SO_2 at 15.00 K; (b) solid SO_2 at 197.64 K (the melting point); (c)
liquid SO_2 at 197.64 K; (d) liquid SO_2 at 263.08 K (the boiling point); and
(e) SO_2 vapor at 263.08 K and 1 atm. With a more accurate evaluation
of the Debye extrapolation and the heat capacity integrals, Giauque and
Stephenson obtained 242.59 $J K^{-1} mol^{-1}$ for the molar entropy of the vapor
at the boiling point and 1 atm.

2-13 Tabulated heat capacity data for gases describe ideal gas behavior at
1 bar. This presents a small obstacle in the calculation of calorimetric standard entropies of gases. The entropy of the *real* gas is determined at the
normal boiling temperature and 1 *atm*, but to complete the calculation with
the tabulated heat capacities the entropy of the *ideal* gas must be calculated
at the boiling temperature and 1 *bar*. Giauque developed a simple procedure
for making this correction beginning with a nonideal gas law,

$$PV_m = RT + \frac{9RPT_c}{128P_c}\left(1 - \frac{6T_c^2}{T^2}\right),$$

called the *Berthelot equation*. This equation of state, and the general entropy
equation,

$$\left(\frac{\partial S_m}{\partial P}\right)_T = -\left(\frac{\partial V_m}{\partial T}\right)_P$$

[see Eq. (2.16)] are used to calculate ΔS_m for a two-step process that
converts a real gas A at the pressure P_1 and temperature T to an ideal gas at
P_2 and T,

$$A \text{ (real gas, } P_1, T) \xrightarrow{1} A \text{ (ideal gas, } 0, T)$$

$$A \text{ (ideal gas, } 0, T) \xrightarrow{2} A \text{ (ideal gas, } P_2, T).$$

Use the Berthelot equation to determine the derivative $(\partial V_m/\partial T)_P$, then integrate the entropy equation for the above two steps, and show that

$$S_A \text{ (ideal gas, } P_2, T) = S_A \text{ (real gas, } P_1, T) + R\ln(P_1/P_2) + \frac{27RT_c^3P_1}{32P_cT^3}.$$

This result is applied in Exercise 2-14, where it is used to calculate the
entropy of an ideal gas at $P_2 = 1$ bar from the entropy of a real gas at
$P_1 = 1$ atm $= 1.0132$ bar.

2-14 The entropy of chlorine gas at the boiling point $T_b = 239.05$ K and 1
atm is 215.22 $J K^{-1}$ (Exercise 2-10). Calculate the standard entropy for chlorine gas at 298.15 K. In the temperature range 100–700 K tabulated standard

heat capacity data for chlorine gas fit the cubic temperature function,

$$C_{P,Cl_2}/(J\,K^{-1}\,mol^{-1}) = 25.74 + 3.937 \times 10^{-2}\,T - 4.655 \times 10^{-5}\,T^2$$
$$+ 1.850 \times 10^{-8}\,T^3.$$

Use the program Entropy, the ideal-gas correction developed in Exercise 2-13 and critical constants listed in Table 1.2.

2-15 Nitrogen undergoes three transitions as it is heated from a low temperature,

$$N_2(sII) \xrightarrow{35.61\ K} N_2(sI) \xrightarrow{63.14\ K} N_2(l) \xrightarrow{77.32\ K} N_2(g).$$

The two solid phases sII and sI are different crystalline modifications. They are treated thermodynamically like the liquid and solid phases. Giauque and Clayton measured enthalpies for these transitions,

$$\Delta_{II,I}H_{N_2} = 228.9\ J\,mol^{-1}$$
$$\Delta_{I,l}H_{N_2} = 720.9\ J\,mol^{-1}$$
$$\Delta_{l,g}H_{N_2} = 5577\ J\,mol^{-1}.$$

They also measured heat capacities; between 15.82 and 35.61 K, their data fit the cubic temperature function

$$C_{P,N_2}(sII, T)/(J\,K^{-1}\,mol^{-1}) = -52.17 + 7.047\,T - 2.377 \times 10^{-1}\,T^2$$
$$+ 3.290 \times 10^{-3}\,T^3.$$

Between 35.61 and 63.14 K, their data fit

$$C_{P,N_2}(sI, T)/(J\,K^{-1}\,mol^{-1}) = 4.509 + 1.598\,T - 2.696 \times 10^{-2}\,T^2$$
$$+ 1.957 \times 10^{-4}\,T^3,$$

and between 63.14 and 77.32 K,

$$C_{P,N_2}(l, T)/(J\,K^{-1}\,mol^{-1}) = 7.652 + 1.803\,T - 2.319 \times 10^{-2}\,T^2$$
$$+ 1.055 \times 10^{-4}\,T^3.$$

Tabulated standard heat capacities for nitrogen gas between 77.32 and 700 K fit the function

$$C_{P,N_2}^0(g, T)/(J\,K^{-1}\,mol^{-1}) = 29.17 - 6.976 \times 10^{-4}\,T + 8.441 \times 10^{-8}\,T^2$$
$$+ 5.982 \times 10^{-9}\,T^3.$$

Calculate the following:

(a) The Debye evaluation of the entropy at 15.82 K using the program Debye. The Debye temperature for nitrogen is 68 K.

(b) Values for the following four heat capacity integrals using the program Entropy

$$\int_{15.82}^{35.61} \frac{C_P(\text{sII}, T)}{T} \, dT$$

$$\int_{35.61}^{63.14} \frac{C_P(\text{sI}, T)}{T} \, dT$$

$$\int_{63.14}^{77.32} \frac{C_P(\text{l}, T)}{T} \, dT$$

$$\int_{77.32}^{298.15} \frac{C_P^0(\text{g}, T)}{T} \, dT.$$

(c) Entropies for the three transitions $N_2(\text{sII}) \rightarrow N_2(\text{sI})$, $N_2(\text{sI}) \rightarrow N_2(\text{l})$, and $N_2(\text{l}) \rightarrow N_2(\text{g})$.

(d) The difference between the entropy of nitrogen as a real gas at 77.32 K and 1 atm and the entropy of nitrogen as an ideal gas at 77.32 K and 1 bar. Use Giauque's method for this correction. (see Exercise 2-13).

(e) The standard entropy of nitrogen at 298.15 K. Giauque and Clayton reported 192.2 $J\,K^{-1}\,mol^{-1}$ at 298.15 K and 1 bar (corrected from the 1 atm pressure condition assumed by Giauque and Clayton).

Ideal Solutions

2-16 Use the program Raoult to calculate and plot real and ideal chemical potentials for acetone in acetone-benzene mixtures at 333.2 K.

2-17 Use the program Raoult to calculate and plot real and ideal chemical potentials for ethanol in ethanol-water mixtures at 298.2 K.

2-18 Use the program Henry to calculate and plot real and ideal chemical potentials for methanol in methanol-water mixtures at 298.2 K.

Nonideal Solutions

2-19 Use the program Vanlaar to calculate the van Laar parameters a and b for benzene (A)–methanol (B) mixtures at 55°C. Relevant vapor pressure data are quoted below. Vapor pressures of pure benzene and methanol at 55°C are 326.94 and 517.65 Torr. Note that x_B and y_B denote mole fractions of methanol in the solution and vapor phases.

x_B	y_B	P/Torr
0.0493	0.4051	527.12
0.1031	0.4841	597.48
0.3297	0.5540	664.24
0.4874	0.5845	675.62
0.4984	0.5858	675.99
0.6076	0.6078	678.44
0.7898	0.6716	664.91
0.9014	0.7697	622.29

2-20 Use the program Vanlaar to calculate the van Laar parameters a and b for acetic acid (A)–pyridine (B) mixtures at 80.05 °C. Relevant vapor pressure data are quoted below. Vapor pressures of pure pyridine and acetic acid at 80.05 °C are 242.75 and 207.60 Torr.

x_B	y_B	P/Torr
0.2434	0.0842	104.0
0.3281	0.2384	88.9
0.3749	0.3482	84.6
0.4442	0.5289	86.4
0.5034	0.6960	94.6
0.5456	0.8054	104.1
0.6186	0.9159	123.5

2-21 Use the program Pitzer to plot mean activity coefficients for the 1-1 electrolytes KOH, NaCl, $NaNO_3$, and $AgNO_3$, the 2-1 electrolytes $MgCl_2$, $CaCl_2$, $BaCl_2$, and the 1-2 electrolytes Na_2SO_4 and Na_2CO_3. What general patterns do these plots follow?

2-22 Use the program Pitzer to plot mean activity coefficients for the 2-2 electrolytes $CdSO_4$, $NiSO_4$, $ZnSO_4$, and $MgSO_4$. How do these plots differ from those seen in the last exercise?

Equilibrium Theory

2-23 Equilibrium constants for reactions involving gas-phase components are independent of pressure, but the equilibrium itself is often influenced by changes in pressure. You can demonstrate this by running the program Gibbs with different pressures entered. Interpret these pressure effects in terms of Le Châtelier's principle. Also run the program with various values of the equilibrium constant K entered and interpret the results.

2-24 Calculate and plot equilibrium constants for the reactions

$$CH_4(g) + H_2O(g) \rightarrow CO(g) + 3\,H_2(g)$$
$$H_2(g) + CO_2(g) \rightarrow CO(g) + H_2O(g)$$

in the temperature range 500 to 2000 K using an adaptation of the program
Vanthoff. Are these reactions endothermic or exothermic?

2-25 In the next chapter we will carry out thermodynamic calculations for
the geochemical reaction

$$CaCO_3(s) + SiO_2(s) \rightarrow CaSiO_3(s) + CO_2(g)$$

taking place at high temperatures and high pressures. You can do the high
temperature part with an adaptation of the program Vanthoff. Use
the program to calculate and plot equilibrium constants in the range 500 to
1000 K.

2-26 Equilibrium constants measured by Larson and Dodge for the
ammonia-synthesis reaction,

$$1/2\ N_2(g) + 3/2\ H_2(g) \rightarrow NH_3(g),$$

at various temperatures are quoted below. Adapt the program Linreg so it
fits the data to Eq. (2.70) and calculates average values of $\Delta_r H^0$ and $\Delta_r S^0$ in
the temperature range covered.

T/°C	K
325	0.0401
350	0.0266
375	0.0181
400	0.0129
425	0.00919
450	0.00659
475	0.00516
500	0.00381

2-27 Data are given below for Bodenstein's measurements of equilibrium
constants for the reaction

$$2\ NO_2(g) \rightarrow 2\ NO(g) + O_2(g)$$

at various temperatures. Use these data and an adaptation of the program
Linreg to estimate $\Delta_r H^0$ and $\Delta_r S^0$ for the reaction.

T/K	$\log_{10} K$
498.9	−4.216
519.5	−3.734
570.4	−2.706
627.5	−1.752
671.4	−1.118
727.7	−0.418
752.7	−0.136
796.3	0.298
825.3	0.570

2-28   Data are given below for measured values of the equilibrium constant for the H_2O_2 formation reaction,

$$H_2(g) + O_2(g) \rightarrow H_2O_2(g),$$

at various temperatures.

T/K	$\ln K$
600	14.66
800	7.615
1000	3.452
1300	-0.493
1500	-2.466

Adapt the program Linreg to fit these data to the equation $\Delta_r G^0 = \Delta_r H^0 - T\Delta_r S^0$ and thus to estimate $\Delta_r H^0$ and $\Delta_r S^0$ in the temperature range covered.

2-29   The reaction

$$N_2O_4(g) \rightarrow 2\,NO_2(g)$$

was studied at equilibrium by Bodenstein and Boës over a range temperatures. Some of their data for equilibrium constants are quoted below. Using these data and an adaptation of the program Linreg estimate $\Delta_r H^0$ and $\Delta_r S^0$ for the reaction.

T/K	$\log_{10} K$	T/K	$\log_{10} K$
281.7	-1.449	334.8	0.254
285.7	-1.273	342.5	0.459
288.8	-1.197	352.0	0.696
293.0	-1.022	361.9	0.903
302.0	-0.729	369.0	1.079
305.9	-0.587	383.1	1.352
315.9	-0.284	390.7	1.543
324.0	-0.036	403.8	1.744

2-30   The H_2S decomposition reaction

$$2\,H_2S(g) \rightarrow 2\,H_2(g) + S_2(g),$$

was studied at equilibrium by Preuner and Schupp. Results for their measurements of equilibrium constants are listed below. Adapt the program Linreg to estimate $\Delta_r H^0$ and $\Delta_r S^0$ for the reaction.

T/K	$K/10^{-5}$
1023	0.89
1103	3.8
1218	24.5
1338	118
1405	260

2-31 Neuman measured the equilibrium solubility of AgCl(s) in solutions containing various concentrations of CaSO$_4$, a strong electrolyte. The added CaSO$_4$ changed the ionic strength s in the solutions. To analyze Neuman's data quoted below first prove that

$$K_s = \gamma_{\pm}^2 c^2,$$

where K_s is the solubility product for AgCl, $\gamma_{\pm}$ is the mean activity coefficient for Ag$^+$ and Cl$^-$, and c is the concentration of AgCl. With Eq. (2.61) calculating $\gamma_{\pm}$,

$$\ln(c^2) = \ln K_s + 2 A\sqrt{s},$$

use the program Linreg to fit Neuman's data to this equation and to calculate a value for K_s

$s^{1/2}/(\text{mol}^{1/2}\,\text{kg}^{-1/2})$	$c/10^{-5}(\text{mol}\,\text{kg}^{-1})$
0.00620	1.281
0.01128	1.287
0.02105	1.306
0.04539	1.344
0.06389	1.372
0.07811	1.395
0.1007	1.436

3
Chemical Thermodynamics in Use

For chemists thermodynamics reduces largely to the simple lesson that the spontaneous direction of a chemical reaction is forward if $\Delta_r G < 0$, backward if $\Delta_r G > 0$, and in equilibrium if $\Delta_r G = 0$. This chapter shows how geochemists, biochemists, and plain chemists apply these conclusions. One approach taken by geochemists, calculation of $\Delta_r G$ as a function of pressure and temperature, is seen in the program `Powell` in Sec 3.1. Chemists most often use thermodynamics to describe chemical reactions in equilibrium. Section 3.2 illustrates with some examples of equilibrium calculations involving gas-phase reactions (programs `Haber` and `Coal`), acid-base titrations (the program `Tcurve`), and buffers (the program `Acetate`). Biochemists rely on $\Delta_r G$ calculations to construct Gibbs energy profiles for entire schemes of biochemical reactions. Some of the special methods of biochemical thermodynamics are illustrated in Sec. 3.3 (programs `Atp1`, `Atp2`, and `Biochem`), and a Gibbs energy profile of the glycolysis series of reactions is plotted by the program `Gprofile`.

3.1   Geochemical Thermodynamics

Geologists who study metamorphic chemical reactions are familiar with a special kind of high-pressure, high-temperature reaction whose components each occur in a separate pure phase. Composition variables are fixed for such a reaction, and that means the reaction Gibbs energy behaves as the function $\Delta_r G(P, T)$; it depends only on P and T.

We obtain the function $\Delta_r G(P, T)$ by first noting that

$$\left(\frac{\partial(\Delta_r G)}{\partial T}\right)_{T, \xi} = \Delta_r V. \tag{3.1}$$

Integrating this equation at a fixed temperature T and extent of reaction ξ, we arrive at

$$\Delta_r G(P, T) = \Delta_r G(1, T) + \int_1^P (\Delta_r V)\, dP', \tag{3.2}$$

with P expressed in bar.

To calculate $\Delta_r G(1, T)$, we write

$$\Delta_r G(1, T) = \Delta_r H(1, T) - T\Delta_r S(1, T) \tag{3.3}$$

[Eq. (2.5)], then recall that

$$\left(\frac{\partial(\Delta_r H)}{\partial T}\right)_{P,\xi} = \Delta_r C_P$$

and

$$\left(\frac{\partial(\Delta_r S)}{\partial T}\right)_{P,\xi} = \frac{\Delta_r C_P}{T}$$

[Eqs. (2.1) and (2.2)], and integrate these equations to calculate $\Delta_r H(1, T)$ and $\Delta_r S(1, T)$,

$$\Delta_r H(1, T) = \Delta_r H(1, 298) + \int_{298}^{T} (\Delta_r C_P)\, dT' \tag{3.4}$$

$$\Delta_r S(1, T) = \Delta_r S(1, 298) + \int_{298}^{T} \frac{\Delta_r C_P}{T'}\, dT'. \tag{3.5}$$

Substituting Eqs. (3.4) and (3.5) in Eq. (3.3) and then in Eq. (3.2), we have one version of the desired function of P and T,

$$\Delta_r G(P, T) = \Delta_r H(1, 298) - T\Delta_r S(1, 298) + \int_{298}^{T} (\Delta_r C_P)\, dT'$$
$$- T\int_{298}^{T} \frac{\Delta_r C_P}{T'}\, dT' + \int_{1}^{P} (\Delta_r V)\, dP'. \tag{3.6}$$

The first two terms on the right in this equation, $\Delta_r H(1, 298)$ and $T\Delta_r S(1, 298)$, are easily calculated from tabulated standard enthalpies of formation and standard entropies for the reaction's components. In the last chapter, we developed methods for calculating the next two terms, the two heat capacity integrals $\int_{298}^{T}(\Delta_r C_P)\, dT'$ and $\int_{298}^{T}(\Delta_r C_P/T')\, dT'$. But the last term, the volume integral $\int_{1}^{P}(\Delta_r V)\, dP'$, is new and complicated for the high-pressure conditions (2000 bar and beyond) we will need to recognize.

Powell has demonstrated that both heat capacity integrals and the volume integral can be simplified for a common kind of metamorphic reaction which forms a gaseous component. An example is the reaction that converts the mineral calcite ($CaCO_3$) to wollastonite ($CaSiO_3$),

$$\text{Calcite} + \text{Quartz} \rightarrow \text{Wollastonite} + CO_2(g).$$
$$CaCO_3 \qquad SiO_2 \qquad\quad CaSiO_3$$

The minerals calcite, quartz, and wollastonite all occur in separate pure solid phases, and (we assume) CO_2 is found by itself in the gas phase. For this reaction

$$\Delta_r V = V_{CO_2} + V_{CaSiO_3} - V_{CaCO_3} - V_{SiO_2}$$
$$= V_{CO_2} + \Delta_r V_s, \qquad (3.7)$$

where $\Delta_r V_s = V_{CaSiO_3} - V_{CaCO_3} - V_{SiO_2}$ is the molar volume change for the solid components alone. This term has been separated because it is nearly independent of pressure and temperature, and its contribution to the volume integral is simply

$$\int_1^P (\Delta_r V_s)\,dP' = (P-1)\Delta_r V_s, \qquad (3.8)$$

with P expressed in bar and $\Delta_r V_s$ in $J\,bar^{-1}\,mol^{-1}$ (to convert $\Delta_r V_s$ in the usual units $L\,mol^{-1}$ to these units note that $L\,mol^{-1} = L\,bar\,bar^{-1}\,mol^{-1} = 100\,J\,bar^{-1}\,mol^{-1}$). Substituting from Eqs. (3.7) and (3.8) in Eq. (3.6), we now have

$$\Delta_r G(P, T) = \Delta_r H(1, 298) - T\Delta_r S(1, 298) + \int_{298}^T (\Delta C_P)\,dT'$$
$$- T\int_{298}^T (\Delta_r C_P/T')\,dT' + (P-1)\Delta_r V_s + \int_1^P V_{CO_2}\,dP'. \qquad (3.9)$$

The remaining problem is to evaluate the volume integral $\int_1^P V_{CO_2}\,dP'$ for the gaseous component. That could be done with a nonideal gas law, but none of the gas laws we have considered is valid in the very high pressure region (2–10 kbar) we are now covering. Powell avoids this problem by taking another tack. His formulation is based on the observations that the two heat capacity integrals are small enough in magnitude to be neglected, and that the CO_2 volume integral in Eq. (3.9) can be expressed as a single function F_{CO_2} of P and T,

$$F_{CO_2} = a' + b'T, \qquad (3.10)$$

where a' and b' are parameters that depend on P and T. Powell provides simple tables of values of a' and b' at the pressures 2, 4, 6, 8 and 10 kbar and in the temperature range 200 to 1000 °C. With these considerable simplifications the Gibbs energy calculation finally reduces to

$$\Delta_r G(P, T) = \Delta_r H(1, 298) - T\Delta_r S(1, 298) + (P-1)\Delta_r V_s + n_{CO_2} F_{CO_2}, \qquad (3.11)$$

in which n_{CO_2} is the stoichiometric coefficient for CO_2 in the reaction ($= 1$ in the reaction considered).

The program `Powell` incorporates Powell's method. It reads from the data file `Chap3.m`: data for enthalpies of formation and entropies of the reaction's components at 1 bar and 298 K; molar volume data for the solid components; and values of a' and b' in Eq. (3.10). It then calculates: $\Delta_r H(1, 298), \Delta_r S(1, 298), \Delta_r V_s, F_{CO_2}$ and finally $\Delta_r G(P, T)$ according to Eq. (3.11). The nect example introduces the program.

Example 3-1. When geologists calculate $\Delta_r G$'s for metamorphic reactions, they usually aim to determine equilibrium conditions, because most of the experimental data are gathered at or near equilibrium. Use the program Powell to calculate the equilibrium temperature for the $CaCO_3$-SiO_2 reaction $P = 2$ kbar. The measured equilibrium temperature at 2 kbar is about 1013 K.

Answer. To run the program Powell enter data for the pressure P, temperature T, identity of the gaseous component (CO_2), the stoichiometric coefficient $n = 1$ for the gaseous component, and enter the reaction in reactionList with products positive and reactants negative,

ReactionList = wollastonite + CO2 - quartz - calcite;

The program calculates $\Delta_r G = 0.04$ kJ mol^{-1} at $P = 2$ kbar and 1018 K, and $\Delta_r G = -0.05$ kJ mol^{-1} at 2 kbar and 1019 K. The calculated equilibrium temperature at 2 kbar is between these temperatures.

The program Powell implements a similar calculation for metamorphic reactions that form the gaseous component $H_2O(g)$. Equation (3.11) is replaced by

$$\Delta_r G(P, T) = \Delta_r H(1, 298) - T\Delta_r S(1, 298) + (P - 1)\Delta_r V_s + n_{H_2O}F_{H_2O},$$
$$\text{(3.12)}$$

with

$$F_{H_2O} = a'' + b''T, \qquad \text{(3.13)}$$

where a'' and b'' are parameters for H_2O, similar to a' and b' for CO_2 in Eq. (3.10). They are also included in the data file Chap3.m.

3.2 Ordinary Equilibrium Calculations

Preliminaries

Most equilibrium calculations are based on equilibrium-constant expressions, which in turn permit activity calculations. For the generic reaction

$$aA + bB \rightarrow rR + sS$$

the equilibrium-constant expression is

$$K = \frac{a_R^r a_S^s}{a_A^a a_B^b},$$ (3.14)

in which the a's represent equilibrium values of activities. One or more of the activities may be unknown, and this equation (usually combined with others) allows us to calculate equilibrium values for the unknown activities.

Remember that an activity for a component A depends on the concentration of A and also perhaps on concentrations of other components in the system. This dependence is expressed in Eqs. (2.39) to (2.42), which we now compress into a single activity equation,

$$a_A = \gamma_A (A),$$ (3.15)

where γ_A is suitable activity coefficient and (A) has various meanings, depending on how A is identified,

$$(A) = p_A/p^0 \text{ for a gas}$$ (3.16)

$$(A) = x_A \text{ for a solvent}$$ (3.17)

$$(A) = m_A/m^0 \text{ for a solute}$$ (3.18)

$$(A) = C_A/C^0 \text{ for a solute.}$$ (3.19)

However it is defined, (A) (like the activity a_A) is a unitless quantity. We will call p_A/p^0 a *unitless partial pressure*, m_A/m^0 a *unitless molality*, and C_A/C^0 a *unitless molarity*. We will also find it convenient to use the standard notation [A] to represent the concentration of a solute A with units included, either a molality or a molarity, or both if the solution is sufficiently dilute.

Varieties of Equilibrium Constants

For special equilibrium calculations some special equilibrium constants are required. In addition to K, defined by Eq. (3.14), we introduce

$$K_C = \frac{(R)^r (S)^s}{(A)^a (B)^b},$$ (3.20)

with the ( )'s representing unitless molalities or molarities, and for gas-phase reactions

$$K_P = \frac{(R)^r (S)^s}{(A)^a (B)^b},$$ (3.21)

with the ( )'s representing unitless partial pressures.

The Ammonia Synthesis Reaction

We take as a first example the gas-phase reaction that synthesizes ammonia from elemental nitrogen and hydrogen,

$$1/2\, N_2(g) + 3/2\, H_2(g) \rightarrow NH_3(g),$$

developed as an industrial process in the early 1900s by Haber and Bosch.

Suppose equilibrium is approached by beginning with a mixture of 1 mol N_2 and 1 mol H_2 and no NH_3. At equilibrium some of the N_2 and H_2 is converted to NH_3 but the total number of *atoms* remains constant. Initially there are 1 mol of N_2 molecules in the system and 2 mol N atoms. If, at some later time, there are n_{N_2} mol N_2 molecules and n_{NH_3} mol NH_3 molecules, the corresponding molar amount of N atoms, $2n_{N_2} + n_{NH_3}$, must still be equal to 2 mol,

$$2n_{N_2} + n_{NH_3} = 2 \text{ mol},$$

because no atoms get lost in the reaction.

We make this equation more convenient by dividing by n, the total molar amount of molecules in the system,

$$\frac{2n_{H_2}}{n} + \frac{n_{NH_3}}{n} = \frac{2 \text{ mol}}{n}.$$

The ratios n_{N_2}/n and n_{NH_3}/n represent y_{N_2} and y_{NH_3}, mole fractions in the gas phase, so we write

$$2y_{N_2} + y_{NH_3} = \frac{2 \text{ mol}}{n},$$

and multiply the equation by P, the total pressure,

$$2y_{N_2}P + y_{NH_3}P = \frac{(2 \text{ mol})P}{n}.$$

The quantities $y_{N_2}P$ and $y_{NH_3}P$ are p_{N_2} and p_{NH_3}, partial pressures of N_2 and NH_3, so we have

$$2p_{N_2} + p_{NH_3} = \frac{(2 \text{ mol})P}{n}. \tag{3.22}$$

This equation conveniently expresses the *nitrogen material balance*.

A similar argument leads to an equation for the *hydrogen material balance*. We begin with 2 mol H atoms (1 mol H_2 molecules) and sometime later have $2n_{H_2} + 3n_{NH_3}$ mol H atoms, with the n's again counting molecules. The total number of moles of H atoms remains constant at 2, so

$$2n_{N_2} + 3n_{NH_3} = 2 \text{ mol},$$

and we convert this to

$$2p_{H_2} + 3p_{NH_3} = \frac{(2 \text{ mol})P}{n}. \tag{3.23}$$

The two material balance equations (3.22) and (3.23) introduce five variables p_{N_2}, p_{H_2}, p_{NH_3}, P and n. To solve the problem we need three more equations. One of these is the equilibrium-constant expression, which we write with gas activities represented as unitless partial pressures (we will set the pressure low enough to justify using values of one for all fugacity coefficients), so

$$K = K_P = \frac{(p_{NH_3}/p^0)}{(p_{N_2}/p^0)^{1/2}(p_{H_2}/p^0)^{3/2}}. \tag{3.24}$$

We also express the total pressure P as a sum of the partial pressures,

$$P = p_{N_2} + p_{H_2} + p_{NH_3}. \tag{3.25}$$

In the following example the equilibrium calculation is done for a total equilibrium pressure of 10 bar and the fifth equation needed to complete the set of equations for the five variables is simply

$$P = 10 \text{ bar}. \tag{3.26}$$

Example 3-2. Use the program Haber to calculate N_2, H_2 and NH_3 partial pressures at 450 °C and 10 bar. At this temperature $K = 0.00659$. Assume that equilibrium is achieved and that the gases behave ideally.

Answer. Enter Eqs. (3.22) to (3.26) in the FindRoot function of Haber as instructed in the program's comment lines. Following the equations enter trial, minimum and maximum values for the five unknown variables p_{N_2}, p_{H_2}, p_{NH_3}, P, and n. Choose the trial values so they are within an order of magnitude of the correct values. In this case that can be 10 bar (or 0.1 bar or 1 bar). Pressures and molar amounts are all positive, so enter zeros for the minimum values. Maximum values place upper limits on the regions in which FindRoot searches for roots. Any large value (e.g., 100 bar) will do. The program calculates $p_{N_2} = 5.000$ bar, $P_{H_2} = 4.843$ bar, and $p_{NH_3} = 0.157$ bar. Under the conditions given the equilibrium ammonia yield is low. As a check, the program calculates the equilibrium constant K from the calculated partial pressures, obtaining the value supplied, $K = 0.00659$.

Gas-phase equilibria are responsive to both pressure and temperature changes. The ammonia synthesis reaction, for example, is exothermic so its equilibrium constant decreases with increasing temperature (Sec. 2.8). If you want a higher yield of ammonia, decrease the temperature, but not too much because lower temperatures decrease the *rate* of the reaction. Le Châtelier's principle predicts that the ammonia yield is also increased by increasing the pressure. Both effects are illustrated by the following results obtained with an adaptation of Haber, assuming that the initial mixture is 1 mol N_2, 3 mol H_2, and 0 mol NH_3 (so there are 6 mol H atoms and 2 mol N atoms):

P/bar	T/K	K	P_{NH_3}/bar
10	700	0.00935	0.2865
10	800	0.00299	0.0953
20	700	0.00935	1.086
20	800	0.00299	0.374

Verify these results for more practice with Haber.

Coal Gasification

The ammonia-synthesis equilibrium involves a single reaction. The principles of the calculation are not changed if more than one reaction contributes, but the calculation becomes more complicated mathematically because more equations must be solved simultaneously. We illustrate with a high-temperature system containing five gaseous components H_2O, CO_2, CO, H_2, and CH_4, and excess of a solid phase containing coal, which we represent as pure graphite or C(s).

One way to approach this equilibrium calculation is to recognize three independent reactions, all heterogeneous,

$$C(s) + 2\,H_2O(g) \xrightarrow{1} CO_2(g) + 2\,H_2(g)$$

$$C(s) + CO_2(g) \xrightarrow{2} 2\,CO(g)$$

$$C(s) + 2\,H_2(g) \xrightarrow{3} CH_4(g).$$

Because the solid phase is pure graphite and we will set the pressure so it is not high, we can assume an activity of one for the solid phase and, as before, calculate gas activities as unitless partial pressures. Equilibrium-constant expressions for the three reactions are

$$K_1 = \frac{(p_{CO_2}/p^0)(p_{H_2}/p^0)^2}{(p_{H_2O}/p^0)^2} = \frac{p_{CO_2}p_{H_2}^2}{p_{H_2O}^2 p^0} \tag{3.27}$$

$$K_2 = \frac{p_{CO}^2}{p_{CO_2}p^0} \tag{3.28}$$

$$K_3 = \frac{p_{CH_4}p^0}{p_{H_2}^2}. \tag{3.29}$$

Suppose the total pressure is 1 bar and the reaction mixture initially comprises excess graphite and 1 mol $H_2O(g)$. Initially we have 2 mol H atoms and this amount remains constant, so at a later time when there are n_{H_2O} mol H_2O molecules, n_{H_2} mol H_2 molecules and n_{CH_4} mol CH_4 molecules there are $2n_{H_2O} + 2n_{H_2} + 4n_{CH_4}$ mol H atoms, and

$$2n_{H_2O} + 2n_{H_2} + 4n_{CH_4} = 2 \text{ mol}.$$

We divide this equation by n, the total number of moles of molecules in the gas phase, and multiply be the total pressure P to obtain

$$2p_{H_2O} + 2p_{H_2} + 4p_{CH_4} = \frac{(2 \text{ mol})P}{n}. \tag{3.30}$$

This is a statement of the hydrogen material balance. Similarly, for the oxygen material balance, we have

$$p_{H_2O} + p_{CO} + 2p_{CO_2} = \frac{(1 \text{ mol})P}{n}. \tag{3.31}$$

We now have five equations in the seven unknowns p_{H_2}, p_{H_2O}, p_{CO}, p_{CO_2}, p_{CH_4}, P, and n. Two more equations are needed; they are

$$p_{H_2} + p_{H_2O} + p_{CO} + p_{CO_2} + p_{CH_4} = P \tag{3.32}$$

for the total pressure, and

$$P = 1 \text{ bar}, \tag{3.33}$$

to assign the given total pressure. In the next example we solve these equations with the program Coal.

Example 3-3. The program Vanthoff calculates the equilibrium constants $K_1 = 3.31, K_2 = 1.70$, and $K_3 = 0.0929$ for reactions 1, 2, and 3 at 1000 K. Use the program Coal to calculate coal gasification partial pressures at 1000 K and 1 bar total pressure.

Answer. The procedure for running Coal is the same as that for Haber. Enter Eqs. (3.27) to (3.33) in the FindRoot function, enter 1 bar for all the trial values, and 0 and 100 bar for minimum and maximum values. Results for 1000 K and two other temperatures are listed below.

T/K	P_{CO_2}/bar	P_{CO}/bar	P_{H_2}/bar	P_{H_2O}/bar	P_{CH_4}/bar
800	0.2279	0.04718	0.2844	0.3313	0.1092
1000	0.07628	0.3601	0.4714	0.07156	0.02064
1200	0.004689	0.4914	0.4935	0.006719	0.003629

Notice that increasing the temperature favors production of CO and H_2. The calculation at 1200 K requires adjustment of the DampingFactor option in

FindRoot, which sets the step sizes taken by FindRoot as it searches for roots. Use 0.01 for this option in the calculation at 1200 K.

In the example we have calculated equilibrium coal gasification partial pressures by assuming that excess graphite is present and that equilibrium is approached through the three heterogeneous reactions 1, 2, and 3. Exactly the same equilibrium state is reached through any other set of three independent reactions collectively involving the graphite solid phase and the five gaseous components H_2O, CO_2, CO, CH_4 and H_2. You can demonstrate this point by making the calculation requested in Exercise 3-8 for the reaction scheme

$$C(s) + 2\,H_2O(g) \xrightarrow{1} CO_2(g) + 2\,H_2(g)$$

$$CO_2(g) + H_2(g) \xrightarrow{4} CO(g) + H_2O(g)$$

$$CO(g) + 3\,H_2O(g) \xrightarrow{5} H_2O(g) + CH_4(g).$$

The first of these reactions is heterogeneous and was recognized in the scheme treated in the example. The other two reactions are homogeneous and different from those seen before.

Buffer Equilibria

In these problems we work with ionization equilibria of weak electrolytes, for example, the ionization of acetic acid (HAc),

$$HAc(aq) \rightleftharpoons H^+(aq) + Ac^-(aq).$$

We designate equilibrium constants for this and other acid ionizations with K_a. In this case, K_a is defined

$$K_a = \frac{a_{H^+} a_{Ac^-}}{a_{HAc}} = 1.754 \times 10^{-5}.$$

which we also express on a logarithmic scale as a pK_a value defined $pK_a = -\log_{10} K_a$. For acetic acid,

$$pK_a = -\log_{10}(1.754 \times 10^{-5})$$

$$= 4.756.$$

See Table 3.1 for a list of pK_a values for some other weak acids. K_{a1}, K_{a2}, and K_{a3} in the table refer to ionization of the first, second, and third hydrogens from a polyprotic acid. For example, the first ionization equilibrium of H_2CO_3, to which K_{a1} applies, is

$$H_2CO_3(aq) \rightleftharpoons HCO_3^-(aq) + H^+(aq),$$

TABLE 3.1. pK_a values for selected weak acids at 25 °C

Acid	pK_{a1}	pK_{a2}	pK_{a3}
Acetic	4.756		
Benzoic	4.212		
Carbonic	6.352	10.329	
Citric	3.128	4.761	6.396
Formic	3.752		
Lactic	3.860		
Oxalic	1.271	4.266	
Phosphoric	2.148	7.198	12.38
Tartaric	3.033	4.366	

Source: H.A. Robinson and R.H. Stokes, 1959, p. 517.

and the second ionization, represented by K_{a2}, is

$$HCO_3^-(aq) \rightleftharpoons H^+(aq) + CO_3^{2-}(aq).$$

We illustrate the method of buffer calculations with an extended example.

Example 3-4. You have prepared an acetate buffer by dissolving 0.0100 mol sodium acetate (NaAc) and 0.0100 mol acetic acid (HAc) in a liter of solution, and you want to calculate the pH in the solution, assuming that unionized acetic acid behaves ideally.

Answer. For this calculation and most others involving electrolyte solutions, we will need to solve simultaneously five kinds of equations:

Equilibrium Constant Expressions

For the acetate buffer these are

$$\frac{\gamma_{H^+}(H^+)\gamma_{Ac^-}(Ac^-)}{(HAc)} = 1.754 \times 10^{-5} \tag{3.34}$$

for acetic acid ionization, and

$$\gamma_{H^+}(H^+)\gamma_{OH^-}(OH^-) = 1.012 \times 10^{-14} \tag{3.35}$$

for water ionization,

$$H_2O \rightleftharpoons H^+(aq) + OH^-(aq).$$

Calculation of Ionic Strength

In the acetate buffer the ions Na$^+$, H$^+$, OH$^-$ and Ac$^-$ contribute to the ionic strength s,

$$s = (1/2)([Na^+] + [H^+] + [OH^-] + [Ac^-]). \tag{3.36}$$

Calculation of Activity Coefficients

We use the approximate Eq. (2.63) for this purpose, with $A = 1.171 \text{ kg}^{1/2}\text{ mol}^{-1/2}$, $b = 3.282 \text{ kg}^{1/2}\text{ mol}^{-1/2}\text{ nm}^{-1}$, and data for γ_B taken from Table 2.5. The three activity coefficient equations for the acetate buffer are

$$\gamma_{H^+} = \exp\left(\frac{(-1.171 \text{ kg}^{1/2}\text{ mol}^{-1/2})\sqrt{s}}{1 + (3.282 \text{ kg}^{1/2}\text{ mol}^{-1/2}\text{ nm}^{-1})(0.90 \text{ nm})\sqrt{s}}\right) \qquad (3.37)$$

$$\gamma_{OH^-} = \exp\left(\frac{(-1.171 \text{ kg}^{1/2}\text{ mol}^{-1/2})\sqrt{s}}{1 + (3.282 \text{ kg}^{1/2}\text{ mol}^{-1/2}\text{ nm}^{-1})(0.35 \text{ nm})\sqrt{s}}\right) \qquad (3.38)$$

$$\gamma_{Ac^-} = \exp\left(\frac{(-1.171 \text{ kg}^{1/2}\text{ mol}^{-1/2})\sqrt{s}}{1 + (3.282 \text{ kg}^{1/2}\text{ mol}^{-1/2}\text{ nm}^{-1})(0.45 \text{ nm})\sqrt{s}}\right). \qquad (3.39)$$

Material Balance Equations

We note that the total acetate unitless concentration is 0.0200 (from both HAc and NaAc),

$$(Ac^-) + (HAc) = 0.0200, \qquad (3.40)$$

and that the Na^+ unitless concentration is 0.0100 (from NaAc, a strong electrolyte),

$$(Na^+) = 0.0100. \qquad (3.41)$$

Statement of Electroneutrality

In all electrolyte solutions the negative charge derived from anions must equal the positive charge from cations. If this condition is not met, the system develops a (possibly very large) static electric charge in which we are not interested. For the acetate buffer, this electroneutrality conditions is

$$(Na^+) + (H^+) = (Ac^-) + (OH^-). \qquad (3.42)$$

This gives us nine equations in the nine unknowns (H^+), (Ac^-), (Hac), (OH^-), (Na^+), γ_{H^+}, γ_{Ac^-}, γ_{OH^-}, and s. For the pH calculation, we add a tenth equation,

$$pH = -\log[\gamma_{H^+}(H^+)], \qquad (3.43)$$

based on the physical chemist's definition of pH as the negative logarithm of the *activity* of H^+.

The program Acetate solves Eqs. (3.34) to (3.43) numerically with the function FindRoot, as in the programs Haber and Coal. Enter the equations first, and then trial values and minimum and maximum values for each variable. Because values of the variables cover a broad range in this problem, trial values cannot all be made the same, as they were in Haber and Coal. Trial values here, and in other buffer problems, need to be obtained in a preliminary, order-of-magnitude calculation.

That is not a difficult task. For the order-of-magnitude calculation we assume that all activity coefficients are equal to one and note that neither the acetic acid nor the acetate ion mixed to make the buffer are much altered in the solution, so we assume that no reaction occurs and calculate

$$(HAc) \cong (Ac^-) \cong 0.01.$$

Then from Eq. (3.34) (setting all activity coefficients equal to one)

$$(H^+) \cong 2 \times 10^{-5},$$

and from Eq. (3.35)

$$(OH^-) \cong 5 \times 10^{-10}.$$

The ionic strength is

$$s \cong (1/2)([H^+] + [Na^+] + [OH^-] + [Ac^-]) \cong 0.01 \text{ mol kg}^{-1},$$

and

$$(Na^+) \cong 0.01.$$

Enter these results for trial values of the variables, and 0 and 1 for the minimum and maximum values. See the program Acetate for further details. The program calculates pH = 4.713 for the buffer, compared to the measured value 4.718.

Titration Equilibria

As an acid is titrated with a base the pH increases during the course of the titration, at first slowly, then rapidly near the equivalence point, and slowly again beyond the equivalence point. We formulate an equilibrium calculation that traces this S-shaped curve and locates the equivalence point.

Suppose you titrate a weak acid HA with a strong base MOH. Unitless molar concentrations for the acid and base are a and b, and the ionization constant for the weak acid is defined by

$$K_{Ca} = \frac{(H^+)(A^-)}{(HA)}.$$

You begin with a volume V_A of the acid, a total volume of $V_0 (\geq V_A)$ and titrate by adding base. The relationship between the volume V_B of base added and the hydrogen concentration (H^+) is

$$V_B = \frac{aV_A f_1 + V_0 K_{Cw}/(H^+) - V_0(H^+)}{b - K_{Cw}/(H^+) + (H^+)}, \tag{3.44}$$

in which

$$f_1 = \frac{K_{Ca}}{K_{Ca} + (H^+)},$$ (3.45)

and

$$K_{Cw} = (H^+)(OH^-).$$

At the equivalence point,

$$aV_A = bV_{B,equiv},$$ (3.46)

and

$$(H^+)_{equiv} = \left(\frac{K_{Cw}K_{Ca}(V_0 + V_{B,equiv})}{aV_A}\right)^{1/2}.$$ (3.47)

The program Tcurve calculates and plots V_B as a function of pH [now defined pH $= -\log_{10}(H^+)$] according to Eqs. (3.44) and (3.45), and calculates $(H^+)_{equiv}$ with Eq. (3.47).

Example 3-5. Use the program Tcurve to calculate the pH at the equivalence point in the titration of 25.00 cm^3 of 0.100 mol L^{-1} acetic acid with 0.100 mol L^{-1} NaOH. Do the calculation first assuming that all activity coefficients are equal to one. Then compare this result with that obtained using Eq. (2.63) to calculate activity coefficients of the ionic components. Assume that the activity coefficient of the unionized acetic acid HAc equals 1.000.

Answer. Obtain $K_a = 10^{-4.756} = 1.754 \times 10^{-5}$ from the pK_a values quoted in Table 3.1. Use $K_w = 1.012 \times 10^{-14}$ for the water ionization equilibrium. If the activity coefficients all have values of one, $K_{Ca} = K_a$ and $K_{Cw} = K_w$. Enter these and other data given in Tcurve and calculate pH $= 8.73$ at the equivalence point. To calculate the activity coefficients note that at the equivalence point 25.00 cm^3 of 0.100 mol L^{-1} HAc have reacted with 25.00 cm^3 of 0.100 mol L^{-1} NaOH to form 50.00 cm^3 of 0.050 mol L^{-1} NaAc. In this case concentrations are low enough that molalities and molarities are equivalent numerically and this concentration can also be expressed 0.050 mol kg^{-1}. Sodium acetate is a strong electrolyte; it forms the ions Na$^+$ and Ac$^-$, and the ionic strength in the NaAc solution is

$$s = (1/2)([Na^+] + [Ac^-])$$
$$= (1/2)(0.050 \text{ mol kg}^{-1} + 0.050 \text{ mol kg}^{-1})$$
$$= 0.050 \text{ mol kg}^{-1}.$$

Use this ionic strength and data from Table 2.5 to calculate activity coefficients for the ions with Eq. (2.63),

$$\gamma_{H^+} = 0.854$$

$$\gamma_{OH^-} = 0.812$$

$$\gamma_{Ac^-} = 0.821.$$

We now need to calculate K_{Ca} and K_{Cw} for use in Tcurve. Since

$$K_a = \frac{a_{H^+} a_{Ac^-}}{a_{HAc}} = \frac{\gamma_{H^+}(H^+)\gamma_{Ac^-}(Ac^-)}{\gamma_{HAc}(HAc)},$$

we have

$$K_{Ca} = \frac{\gamma_{HAc}}{\gamma_{H^+}\gamma_{Ac^-}} K_a$$

$$= \frac{(1.000)}{(0.854)(0.821)}(1.754 \times 10^{-5})$$

$$= 2.502 \times 10^{-5},$$

and

$$K_{Cw} = (H^+)(OH^-)$$

$$= K_w/\gamma_{H^+}\gamma_{OH^-}$$

$$= \frac{(1.012 \times 10^{-14})}{(0.854)(0.812)}$$

$$= 1.459 \times 10^{-14}.$$

Enter these data in Tcurve and calculate pH = 8.57 at the equivalence
point. The difference between the two pH calculations is 2% in pH and 31%
in (H^+). The activity coefficients do make a difference.

The calculation done by Tcurve concerns titration of a weak acid with a
strong base. In Exercise 3-14 another program is introduced that makes a
similar calculation for titration of a weak acid and a *weak* base.

3.3   Biochemical Thermodynamics

So far in our treatment of chemical thermodynamics, we have been able to
express nonideality of solute components by calculating suitable activity coefficients, at least approximately. For many biochemical components that is
not feasible.

A good example is adenosine triphosphate, often included in biochemical reactions which would, without adenosine triphosphate participation, proceed in the wrong (i.e., biochemically nonbeneficial) direction. In aqueous solutions, adenosine triphosphate can take a series of acidic forms which we will write as $ATP^{4-}, HATP^{3-}, H_2ATP^{2-}$, etc. It also binds Mg^{2+} in biological cellular environments to form $MgATP^{2-}$, $MgHATP^-$, etc. Biochemists usually abandon the task of recognizing all of these distinct components and instead just write ATP to represent *all* of the different forms. We will need to elaborate this notation a bit and write "ATP" when this is intended, so we do not have to contend with conflicting notations, such as ATP and ATP^{4-}.

This point of view demands use of the total "ATP" concentration $(ATP)_T$, defined

$$(ATP)_T = (ATP^{4-}) + (HATP^{3-}) + (MgATP^{2-}) + (MgHATP^{2-}) + \cdots,$$

with the ( )'s referring to unitless molar concentrations. We will also need to define a total activity coefficient γ_{ATP} relative to the total concentration $(ATP)_T$,

$$a_{ATP} = \gamma_{ATP}(ATP)_T. \tag{3.48}$$

Then the "ATP" chemical potential equation is

$$\mu_{ATP} = \mu^0_{ATP} + RT \ln \gamma_{ATP} + RT \ln(ATP)_T$$
$$= \mu^{0\prime}_{ATP} + RT \ln(ATP)_T, \tag{3.49}$$

in which $\mu^{0\prime}_{ATP} = \mu^0_{ATP} + RT \ln \gamma_{ATP}$. This new standard chemical potential incorporates the activity coefficient γ_{ATP} and therefore depends not only on pressure and temperature, but also on composition variables such as (H^+), (Mg^{2+}), and ionic strength.

This approach revises the general chemical potential equation (2.38) for a component A,

$$\mu_A = \mu^{0\prime}_A + RT \ln(A)_T. \tag{3.50}$$

From this, and similar equations for other components, we calculate the Gibbs energy for the generic reaction,

$$aA + bB \rightarrow rR + sS,$$

obtaining

$$\Delta_r G = \Delta_r G^{0\prime} + RT \ln \frac{(R)^r_T (S)^s_T}{(A)^a_T (B)^b_T}, \tag{3.51}$$

with

$$\Delta_r G^{0\prime} = r\mu^{0\prime}_R + s\mu^{0\prime}_S - a\mu^{0\prime}_A - b\mu^{0\prime}_B,$$

called an *apparent standard Gibbs energy*. We will also write Eq. (3.51)

$$\Delta_r G = \Delta_r G^{0\prime} + RT \ln Q', \tag{3.52}$$

in which

$$Q' = \frac{(\mathrm{R})_T^r (\mathrm{S})_T^s}{(\mathrm{A})_T^a (\mathrm{B})_T^b} \tag{3.53}$$

is the reaction's *concentration quotient*. At equilibrium $\Delta_r G = 0$, Q' has an equilibrium value, which we write K' and call an *apparent equilibrium constant*. Equation (3.52) becomes

$$\Delta_r G^{0\prime} = -RT \ln K'. \tag{3.54}$$

Eqs. (3.52) and (3.54) are the biochemist's analogs of Eqs. (2.43) and (2.44).

The apparent equilibrium constant K' is similar to equilibrium constants of the K_C kind we have used before (Sec. 3.2). Both K' and K_C omit activity coefficients. The main difference is that K' expressions are written with total concentrations, while K_C expressions include individual components. Biochemists also have the habit of omitting from K' expressions concentration factors for H^+ and all complexing cations (e.g., Mg^{2+}). Thus for the "ATP" hydrolysis,

$$\text{"ATP"} + H_2O \rightarrow \text{"ADP"} + \text{"P"},$$

where "ADP" represents all the adenosine diphosphate components and "P" all the phosphate components, we have the simple statement

$$K' = \frac{(\mathrm{ADP})_T (\mathrm{P})_T}{(\mathrm{ATP})_T}.$$

A typical K_C expression for one of the many equilibria involved in the "ATP" hydrolysis is

$$K_C = \frac{(\mathrm{MgADP}^-)(\mathrm{H_2P}^-)}{(\mathrm{MgATP}^{2-})},$$

in which the individual components $MgADP^-$ and H_2P^- and $MgATP^{2-}$ are recognized. Another important difference between K' and K_C equilibrium constants is that the former depends on composition, but the latter does not, except for a slight dependence on ionic strength.

We now have the equipment to calculate the Gibbs energy for a biochemical reaction. The first step is to obtain a suitable value of $\Delta_r G^{0\prime}$. The second is to supply the total concentrations appearing in the concentration quotient Q' of Eq. (3.53) for the biological situation of interest. And in the third the two results, $\Delta_r G^{0\prime}$ and Q', are combined in Eq. (3.52) to calculate $\Delta_r G$.

From the calculational point of view the most difficult part of this procedure is the first step, determination of $\Delta_r G^{0\prime}$ [or K' if Eq. (3.54) is used], which is complicated because $\Delta_r G^{0\prime}$ depends on composition variables, particularly pH. We next illustrate this detailed calculation with some examples, simple and complicated.

A Simple Example

Consider the biochemical reaction,

$$\text{``F''} \rightarrow \text{``M''},$$

where "F" and "M" represent all of the components derived from fumaric and maleic acids. Both acids are diprotic: fumaric acid forms the components H_2F, HF^-, and F^{2-} and maleic acid forms H_2M, HM^-, and M^{2-}.

Fumaric acid has two ionization equilibria

$$H_2F \rightleftharpoons HF^- + H^+$$

$$HF^- \rightleftharpoons F^{2-} + H^+,$$

for which we write two equilibrium-constant expressions

$$K_{11} = \frac{(HF^-)(H^+)}{(H_2F)}$$

$$K_{12} = \frac{(F^{2-})(H^+)}{(HF^-)}.$$

From these two equations, we derive

$$(HF^-) = \frac{(F^{2-})(H^+)}{K_{12}},$$

$$(H_2F) = \frac{(F^{2-})(H^+)^2}{K_{11}K_{12}},$$

and

$$(F)_T = (H_2F) + (HF^-) + (F^{2-})$$

$$= (F^{2-})\left(1 + \frac{(H^+)}{K_{12}} + \frac{(H^+)^2}{K_{11}K_{12}}\right). \tag{3.55}$$

For maleic acid, we also recognize two ionization equilibria,

$$H_2M \rightleftharpoons HM^- + H^+$$

$$HM^- \rightleftharpoons M^{2-} + H^+,$$

two equilibrium-constant expressions,

$$(HM^-) = \frac{(M^{2-})(H^+)}{K_{22}},$$

$$(H_2M) = \frac{(M^{2-})(H^+)^2}{K_{21}K_{22}},$$

and derive

$$(M)_T = (M^{2-})\left(1 + \frac{(H^+)}{K_{22}} + \frac{(H^+)^2}{K_{21}K_{22}}\right). \tag{3.56}$$

Equations (3.55) and (3.56) permit us to express the apparent equilibrium constant K' as a function of (H^+). We have

$$
\begin{aligned}
K' &= \frac{(M)_T}{(F)_T} \\
&= \frac{(M^{2-})[1 + (H^+)/K_{22} + (H^+)^2/K_{21}K_{22}]}{(F^{2-})[1 + (H^+)/K_{12} + (H^+)^2/K_{11}K_{12}]}.
\end{aligned}
\tag{3.57}
$$

The factor $(M^{2-})/(F^{2-})$ on the right side of this equation represents an equilibrium constant of the K_C kind for the equilibrium

$$F^{2-} \rightleftharpoons M^{2-},$$

one of the many equilibria involved in the system. We label this equilibrium constant K_{ref},

$$K_{\text{ref}} = \frac{(M^{2-})}{(F^{2-})}, \tag{3.58}$$

and note that it, like other K_C equilibrium constants, is independent of pH. Equation (3.57) now shows explicitly how K' depends on (H^+),

$$K' = K_{\text{ref}} \frac{[1 + (H^+)/K_{22} + (H^+)^2/K_{21}K_{22}]}{[1 + (H^+)/K_{12} + (H^+)^2/K_{11}K_{12}]}. \tag{3.59}$$

This equation requires four equilibrium constants, K_{11} and K_{12} for fumaric acid, and K_{21} and K_{22} for maleic acid. Their values are listed (as pK_C values) in Table 3.2, along with similar data for some other biochemical components. A value for K_{ref} is obtained either by direct measurement or by measuring K' at a specific pH and calculating K_{ref} with Eq. (3.59).

The program Biochem uses Eqs. (3.58) and (3.59) to calculate and plot $\Delta_r G^{0\prime}$ at various pH's for the "F" $\rightarrow$ "M" reaction. Similar treatments of several other biochemical reactions are covered in Exercises 3.16 to 3.18.

TABLE 3.2. Ionization constants for some biochemical components at 25 °C

Component	pK_{C1}	pK_{C2}	pK_{C3}
Phosphoric acid	2.15	6.82	12.38
Fumaric acid	3.02	4.40	—
Maleic acid	3.48	5.11	—
Pyruvic acid	2.39	—	—
Lactic acid	3.73	—	—
Citric acid	3.13	4.76	6.40
Oxaloacetic acid	2.22	3.89	—
Succinic acid	4.21	5.72	—
Glucose 6-phosphate	0.94	6.11	—
Fructose 6-phosphate	0.97	6.11	—
Fructose 1,6-diphosphate	1.48	6.29	—
Glyceraldehyde 3-phosphate	1.45	6.45	—
Benzoyltyrosine	3.70	—	—
Glycinamide	7.93	—	—

Sources: R.M. Dawson et. al, 1986; J.T. Edsall and H. Gutfreund, 1983.

A Complicated Example

One of the most important of all biochemical reactions, the "ATP" hydrolysis,

$$\text{"ATP"} + H_2O \rightarrow \text{"ADP"} + \text{"P"},$$

is unfortunately also one of the most complicated. The source of the complexity, as in the previous examples, is the habit of the reaction's components to exist in more than one form. The list of possible forms is a long one in this case, including not only protonated forms of the kind seen before, but also complexes resulting from the binding of adenosine triphosphate, adenosine diphosphate, and phosphate with Mg^{2+} and Ca^{2+}. Under conditions likely to be important biochemically [with $(Mg^{2+}) \gg (Ca^{2+})$], adenosine triphosphate not only forms the bare anion ATP^{4-}, but also the protonated $HATP^{3-}$ and the complexed $MgHATP^-$. The lists for adenosine diphosphate and phosphate are similar, including ADP^{3-}, $HADP^{2-}$, $MgADP^-$ $MgHADP$, HP^-, $MgHP$, and MgH_2P^+.

Calculation of $\Delta_r G^{0\prime}$ is now much more complicated, not only because the components have so many forms, but also because at least two composition variables, pH and pMg $(= -\log[Mg^{2+}])$, need to be considered. For accurate calculations a third variable, the ionic strength s, is added.

We lack the space to describe this calculation further. Instead we offer two programs, Atp1 and Atp2, which demonstrate the complicated dependence of $\Delta_r G^{0\prime}$ for the "ATP" hydroysis on pH, pMg and s.

Gibbs Energy Profiles

If $\Delta_r G^{0\prime}$ values are available for a series of biochemical reactions, the next step in the thermodynamic analysis of the reactions, calculation of Gibbs reaction energies $\Delta_r G$ according to Eq. (3.52), can be taken. This calculation requires total concentrations of all the components in the biological situation of interest. To illustrate, a calculation of this kind for the glycolysis series of reactions is displayed in the program Gprofile. Run the program and note that most of the reactions in the series have Gibbs energies of small magnitudes indicating that these reactions are nearly reversible. Three of the reactions display large negative Gibbs reaction energies. They are important in the biochemical control of the reaction.

3.4   Exercises

Geochemical Thermodynamics

3-1   Calculate an equilibrium temperature for the metamorphic reaction

$$\text{Muscovite} \quad + \text{Quartz} \rightarrow \text{Sillimanite} + \text{Sanidine} + H_2O(g)$$
$$KAl_3Si_3O_{10}(OH)_2 \qquad SiO_2 \qquad Al_2SiO_5 \qquad KAlSi_3O_8$$

at the pressure 2 kbar. The measured equilibrium temperature for this reaction at 2 kbar is about 883 K. What mineral phases are stable above the equilibrium temperature, and what phases are stable below the equilibrium temperature?

3-2   Calculate equilibrium temperatures for the metamorphic reaction

$$\text{Calcite} + \text{Quartz} \rightarrow \text{Wollastonite} + CO_2(g)$$
$$CaCO_3 \qquad SiO_2 \qquad CaSiO_3$$

at 2, 4, 6, and 8 kbar. Enter these results in the program Linreg, fit the equilibrium P data to a quadratic function of T and plot the curve. This kind of plot is a *phase diagram* with the curve representing equilibrium conditions. On the high-temperature side of the equilibrium curve wollastonite and CO_2 are stable, and on the low-temperature side calcite and quartz.

3-3   Calculate an equilibrium temperature for the metamorphic reaction

$$\text{Grossular} \quad + \text{Quartz} \rightarrow \text{Anorthite} \quad + 2\ \text{Wollastonite}$$
$$Ca_3Al_2Si_3O_{12} \qquad SiO_2 \qquad CaAl_2Si_2O_8 \qquad CaSiO_3$$

at 5 kbar. Notice that this reaction does not have a gaseous component. The measured equilibrium temperature at 5 kbar is about 990 K.

3-4 Calculate an equilibrium temperature for the reaction

$$\text{Brucite} \quad \rightarrow \text{Periclase} + H_2O(g)$$
$$Mg(OH)_2 \qquad MgO$$

at 2 kbar. The measured equilibrium temperature at 2 kbar is about 928 K.

3-5 Calculate an equilibrium temperature for the reaction

$$\text{Magnesite} \rightarrow \text{Periclase} + CO_2(g)$$
$$MgCO_3 \qquad MgO$$

at 2 kbar. The measured equilibrium temperature at 2 kbar is about 1118 K.

3-6 Calculate an equilibrium temperature for the metamorphic reaction

$$\text{Talc} \qquad \rightarrow 3\text{Enstatite} + \text{Quartz} + H_2O(g)$$
$$Mg_3Si_4O_{10}(OH)_2 \qquad MgSiO_3 \qquad SiO_2$$

at 2 kbar. The measured value for the equilibrium temperature at 2 kbar is about 1008 K.

Ordinary Equilibrium Calculations

3-7 Write an adaptation of the program `Haber` which makes the ammonia-synthesis calculation (see Example 3-2) with the volume given rather than the pressure. Use the program to make the calculation for $V = 10.0$ L, $T = 700$ K, $K = 0.00935$, and an initial mixture comprising 1 mol N_2, 3 mol H_2, and no NH_3. Assume that all components are ideal.

3-8 Consider three of the gas-phase reactions mentioned in Sec. 3.2,

$$C(s) + 2 H_2O(g) \xrightarrow{\ 1\ } CO_2(g) + 2 H_2(g)$$

$$CO_2(g) + H_2(g) \xrightarrow{\ 4\ } CO(g) + H_2O(g)$$

$$CO(g) + 3 H_2(g) \xrightarrow{\ 5\ } H_2O(g) + CH_4(g).$$

Equilibrium constants for these reactions (calculated by the program `Vanthoff`) are $K_1 = 3.31$, $K_4 = 0.717$, and $K_5 = 0.0391$ at 1000 K. Write an adaptation of the program `Coal` (Example 3-3), which calculates all of the equilibrium partial pressures in this system at 1000 K, with the total pressure $P = 1.00$ bar and beginning with 1 mol $H_2O(g)$. Compare results from this program with those obtained in Example 3-3. Assume that all components are ideal.

3-9 Write an adaptation of the program `Coal` to calculate equilibrium partial pressures at 1000 K for the components involved in the reactions

$$CH_4(g) + H_2O(g) \xrightarrow{\ 6\ } CO(g) + 3 H_2(g)$$

$$CO(g) + H_2O(g) \xrightarrow{\ 7\ } H_2(g) + CO_2(g),$$

whose equilibrium constants at 1000 K are $K_6 = 26.601$ and $K_7 = 1.439$. The initial reaction mixture comprises 2 mol CH_4 and 3 mol H_2O. The volume of the system is 60.0 L. Assume that all components are ideal.

3-10 You have decided to synthesize methanol from carbon monoxide and hydrogen according to the reaction

$$CO(g) + 2H_2(g) \xrightarrow{8} CH_3OH(g),$$

whose equilibrium constant at 500 K is 6.0×10^{-3}. The process is carried out in a 10.0 L batch chemical reactor at 500 K, and the initial reaction mixture consists of 1 mol CO and 2 mol H_2. Write an adaptation of the program Coal to calculate partial pressures and the total pressure at equilibrium for this system. Also calculate the initial pressure in the reactor. Assume that all components are ideal.

3-11 Write an adaptation of the program Acetate (see Example 3-4), which calculates the pH in a carbonate buffer prepared by equilibrating 0.0100 mol Na_2CO_3 and 0.0100 mol $NaHCO_3$ in a liter of solution. The carbonate equilibrium involved is

$$HCO_3(aq)^- \rightleftharpoons H^+(aq) + CO_3^{2-}(aq),$$

whose equilibrium constant at 25 °C is $K = 4.72 \times 10^{-11}$. Calculate activity coefficients with Eq. (2.63) and data from Table 2.5. To estimate initial values for the concentrations you can assume that CO_3^{2-} and HCO_3^- do not react appreciably, so $(CO_3^{2-}) \cong 0.0100$ and $(HCO_3^-) \cong 0.0100$. The measured pH is 10.112.

3-12 A solution prepared by dissolving 0.0500 mol citric acid in a liter of solution is sometimes used as a low-pH buffer. Citric acid has three ionizable hydrogens. The three equilibria that dissociate these hydrogens are ($C = $ Citrate)

$$H_3C(aq) \rightleftharpoons H_2C^-(aq) + H^+(aq)$$

$$H_2C^-(aq) \rightleftharpoons HC^{2-}(aq) + H^+(aq)$$

$$HC^{2-}(aq) \rightleftharpoons C^{3-}(aq) + H^+(aq),$$

with equilibrium constants K_1, K_2 and K_3, whose values are listed (as pK's) in Table 3.1. Write a program as an adaptation of the program Acetate, which calculates the pH in the citrate buffer described. To estimate initial values of the concentrations you can assume that the first ionization dominates and that $(H^+) \cong (H_2C^-)$. The measured pH in this buffer is 2.238.

3-13 Use the program Tcurve to calculate the pH at the equivalence point when 25.0 cm³ of 0.100 mol L^{-1} benzoic acid is titrated with 25.0 cm³ of 0.100 mol L^{-1} NaOH. Calculate activity coefficients with Eq. (2.63) and data from Table 2.5. Assume that at the equivalence point the system consists of a

0.05 $mol\,L^{-1}$ solution of sodium benzoate in which the ionic strength is 0.050 $mol\,kg^{-1}$.

3-14 Equation (3.44) expresses the dependence of the volume V_B of base in the titration of a weak acid with a *strong* base. The corresponding equation for titration of a weak acid HA with a *weak* base MOH is

$$V = \frac{aV_A f_1 + V_0 K_{Cw}/(H^+) - V_0(H^+)}{b f_2 - K_{Cw}/(H^+) + (H^+)}. \tag{3.60}$$

The two functions f_1 and f_2 are

$$f_1 = \frac{K_{Ca}}{K_{Ca} + (H^+)}, \tag{3.61}$$

as before, and

$$f_2 = \frac{K_{Cb}}{K_{Cb} + K_{Cw}/(H^+)}, \tag{3.62}$$

where K_{Ca} and K_{Cb} are ionizations constants for HA and MOH,

$$K_{Ca} = \frac{(H^+)(A^-)}{(HA)}$$

$$K_{Cb} = \frac{(M^+)(OH^-)}{(MOH)}.$$

At the equivalence point (H^+) is calculated with

$$(H^+)_{equiv} = \left(\frac{K_{Ca} K_{Cw}}{K_{Cb}}\right)^{1/2}. \tag{3.63}$$

The program $Ex3-14$ plots titration curves according to Eq. (3.60) and calculates the pH at the equivalence point according to Eq. (3.63). Run $Ex3-14$ with $K_{Ca} = 10^{-4}$, $K_{Cb} = 10^{-5}$, then with $K_{Ca} = 10^{-5}$, $K_{Cb} = 10^{-5}$, and finally with $K_{Ca} = 10^{-5}$, $K_{Cb} = 10^{-4}$. Assume anything for a and b. How does the pH at the equivalence point reflect a comparison of K_{Ca} and K_{Cb}?

Biochemical Thermodynamics

3-15 The biochemical components fructose 1,6-diphosphate and glyceraldehyde 3-phosphate both behave as diprotic acids. Their ionization equilibria are

$$H_2FDP \rightleftharpoons H^+ + HFDP^-$$

$$HFDP^- \rightleftharpoons H^+ + FDP^{2-},$$

with equilibrium constants K_{11} and K_{12}, and

$$H_2GAP \rightleftharpoons H^+ + HGAP^-$$

$$HGAP^- \rightleftharpoons H^+ + GAP^{2-},$$

with equilibrium constants K_{21} and K_{22}. Consider the following reaction found in the glycolysis series of reactions,

$$\text{"FDP"} \rightarrow 2\text{"GAP"},$$

with "FDP" and "GAP" representing all acidic forms of fructose 1,6-diphosphate and glyceraldehyde 3-phosphate. Derive equations that relate: (a) $(FDP)_T$ to (FDP^{2-}) and (H^+); (b) $(GAP)_T$ to (GAP^{2-}) and (H^+); (c) K' for the reaction to (H^+) and $K_{ref} = (GAP^{2-})/(FDP^{2-})$; and (d) the reaction's $\Delta_r G^{0\prime}$ to K'. Note that $K' = 3.78 \times 10^{-6}$ when pH = 7.00. Also derive equations for K' at the extremes of low and high pH. The program Ex3-15 implements all of these calculations. Run the program and characterize general features of the dependence of $\Delta_r G^{0\prime}$ on pH.

3-16   Consider another reaction from the glycolysis scheme,

$$G + \text{"P"} \rightarrow \text{"G6P"},$$

in which "P" and "G6P" represent all acidic forms of phosphate and glucose 6-phosphate. Glucose 6-phosphate behaves as a diprotic acid and phosphate as a triprotic acid. Relevant ionization equilibria are

$$H_3P \rightleftharpoons H_2P^- + H^+$$

$$H_2P^- \rightleftharpoons HP^{2-} + H^+$$

$$HP^{2-} \rightleftharpoons P^{3-} + H^+,$$

with equilibrium constants K_{11}, K_{12}, and K_{13}, and

$$H_2G6P \rightleftharpoons HG6P^- + H^+$$

$$HG6P^- \rightleftharpoons G6P^{2-} + H^+,$$

whose equilibrium constants are K_{21} and K_{22}. Glucose is unaffected by changes in pH. Derive equations that relate: (a) $(P)_T$ to (P^{3-}) and (H^+); (b) $(G6P)_T$ to $(G6P^{2-})$, (H^+) and $K_{ref} = (G6P^{2-})/(P^{3-})(G)$; and (c) the reactions $\Delta_r G^{0\prime}$ to K'. Note that $K' = 0.00382$ at pH = 7.00. Also derive equations for K' related to (H^+) at the extremes of low and high pH. These calculations are all implemented in the program Ex3-16. Run the program and characterize general features of the dependence of $\Delta_r G^{0\prime}$ on pH.

3-17   The components benzoyltyrosine and glycinamide react to form benzoyltyrosyl-glycylamide,

$$\text{"BT"} + \text{"GA"} \rightarrow BTGA + H_2O,$$

with "BT" and "GA" representing all acidic forms of benzoyltyrosine and glycinamide. Both of these components behave as monoprotic acids. Their ionization equilibria are

$$HBT \rightleftharpoons BT^- + H^+$$

$$H_3GA^+ \rightleftharpoons H_2GA + H^+,$$

with ionization constants K_1 and K_2. BTGA is not influenced by changes in pH. Derive an equation that relates the reaction's K' to (H^+) and

$$K_{ref} = \frac{(BTGA)}{(BT^-)(H_3GA^+)}.$$

These calculations are implemented in the program Ex3-17. Run the program and characterize general features of the dependence of $\Delta_r G^{0\prime}$ on (H^+).

3-18 The programs ATP1 and ATP2 calculate and plot $\Delta_r G^{0\prime}$ for the adenosine triphosphate hydrolysis reaction over a range of pH's and pMg's and a given ionic strength. Run both programs and note the complexities of the plots. Use one of the programs to calculate $\Delta_r G^{0\prime}$ for pH = 6.8, pMg = 3.0, and $s = 0.15$ mol kg^{-1}. These values are typical for cellular conditions.

4
Quantum Theory

Physical chemists approach their problems from two broadly different points of view. The first three chapters emphasized gas laws and chemical thermodynamics and adopted a *macroscopic* viewpoint; the systems described were of ordinary size, comprising roughly Avogadro's number of molecules. The next topic, quantum theory, explores the completely different *microscopic* realm of atoms and molecules.

The master equation for this effort is the *Schrödinger equation* introduced in Sec. 4.1. Methods for solving the Schrödinger equation numerically are introduced in Sec. 4.2 (programs Schroed1, Schroed2, Schroed3, and Well). Then the broad problem of describing the motion of electrons and nuclei in atoms and molecules is taken up. Topics included are the orbital approximation in Sec. 4.3 (programs Duality, Aorbital, and Morbital); hierarchies of different kinds of atomic and molecular motion and energy levels in Secs. 4.4 and 4.5 (programs Abc, Morse, Ro1, Ro2, Rovi1, and Vie11). In Sec. 4.6 two programs (Hueckel and Hartree) illustrate some of the matrix methods for molecular orbital calculations.

4.1   Schrödinger Equations

The simplest thing you can say about the Schrödinger equation is that it is always an energy equation, representing the statement

$$\text{Kinetic Energy} + \text{Potential Energy} = \text{Total Energy}. \qquad (4.1)$$

The equation can take different mathematical forms. In our problems we will see it as a differential equation and also as a matrix equation.

We look first at the Schrödinger equation applied to a vibrating diatomic molecule (Fig. 4.1). The bond connecting the two atoms stretches and compresses by the amount x. If x is small, the potential energy for the vibrating molecule is calculated approximately with

$$V(x) = kx^2/2, \qquad (4.2)$$

69

(a)

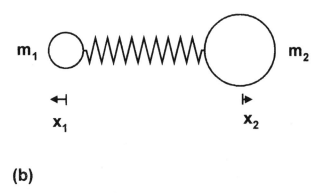

(b)

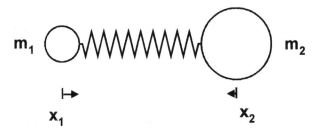

FIGURE 4.1. Illustrating the vibration of a diatomic molecule (a) outward and (b) inward. The variable x in Eqs. (4.2) and (4.3) is related to the two displacements x_1 and x_2 according to $x = x_1 + x_2$.

in which k is force constant. A vibrating molecule behaving this way is called an *harmonic oscillator*. A more realistic potential energy function, approximately valid for any value of x, is the *Morse function*,

$$V(x) = D_e(1 - e^{-\beta x})^2, \tag{4.3}$$

where D_e and β are constants. A vibrating molecule modeled with the Morse function is a *Morse anharmonic oscillator*.

The two potential energy functions (4.2) and (4.3) are compared in Figures 4.2 and 4.3. It is seen that for large positive values of x potential energy calculated according Eq. (4.3) approaches D_e. In this way dissociation of the molecule into atoms is simulated, with D_e representing the dissociation energy.

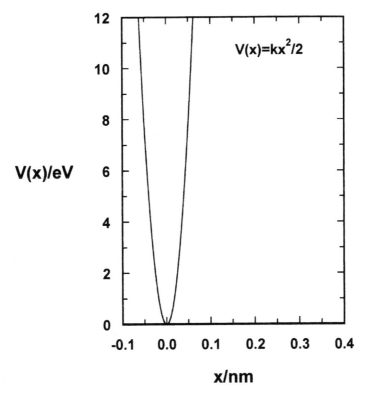

FIGURE 4.2. Plot of the potential energy function $V(x) = kx^2/2$ with $k = 1000 \, \mathrm{J\,m^{-2}}$.

The Schrödinger equation as a differential equation for these cases is

$$-\frac{\hbar^2}{2\mu}\frac{d^2\psi}{dx^2} + V(x)\psi = E\psi, \qquad (4.4)$$

where ψ is a wave function, μ is the diatomic molecule's reduced mass,

$$\mu = \frac{m_1 m_2}{m_1 + m_2},$$

$\hbar = h/2\pi$, E is the total (kinetic + potential) energy, and $V(x)$ is given by Eq. (4.2) for the harmonic oscillator and by Eq. (4.3) for the Morse anharmonic oscillator. The first term on the left in Eq. (4.4) represents kinetic energy, the second potential energy and the term on the right total energy. In this sense the Schrödinger equation is an energy equation. As we will confirm in the next section, solution of the harmonic oscillator version of Eq. (4.4) results in an equation for the total energy of the vibrating diatomic molecule,

$$E = h\omega_e(v + \tfrac{1}{2}), \qquad (4.5)$$

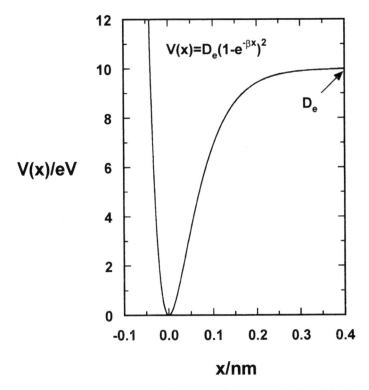

FIGURE 4.3. Plot of the Morse potential energy function with $k = 1000 \ \mathrm{J\,m^{-2}}$ and $D_e = 10$ eV.

where v is a vibrational quantum number with zero and integer values, $v = 0, 1, 2, \ldots$, and ω_e is related to the force constant k and reduced mass μ,

$$\omega_e = \frac{1}{2\pi} \left(\frac{k}{\mu} \right)^{1/2}. \tag{4.6}$$

By simply changing the potential energy factor $V(x)$ in Eq. (4.4), and perhaps the mass factor [μ in Eq. (4.4)], the equation can be applied to any other one-dimensional problem. Consider, for another example, an electron whose mass is m_e, confined to a one-dimensional "well" defined by the potential energy function

$$
\begin{aligned}
V(x) &= V_0 \quad \text{for } x < 0 \\
&= 0 \quad \text{for } 0 \leq x \leq a \\
&= V_0 \quad \text{for } x > a,
\end{aligned} \tag{4.7}
$$

in which a is the width of the well and V_0 the depth. The Schrödinger equation for this case is obtained from Eq. (4.4) by replacing μ with m_e and defining $V(x)$ as in Eqs. (4.7),

$$-\frac{\hbar^2}{2m_e}\frac{d^2\psi}{dx^2} + V(x)\psi = E\psi. \qquad (4.8)$$

4.2 The Schrödinger Equation Solved

For the Harmonic Oscillator

An approximate numerical method, implemented by the program Schroedl, solves (i.e., integrates) the harmonic oscillator Schrödinger equation, that is, Eq. (4.4) with $V(x)$ evaluated by Eq. (4.2).

The equation solved by Schroedl is a reduced version of Eq. (4.4), prepared by transforming the independent variable x to a unitless variable x', with

$$x = \frac{\hbar^{1/2}}{(\mu k)^{1/4}}\,x',$$

or

$$x^2 = \frac{\hbar}{(\mu k)^{1/2}}\,x'^2,$$

and

$$dx^2 = \frac{\hbar}{(\mu k)^{1/2}}\,dx'^2.$$

With this switch in variables, Eq. (4.4) conveniently loses all its constant factors and becomes

$$\frac{d^2\psi}{dx'^2} = (x'^2 - E')\psi, \qquad (4.9)$$

where

$$E' = \frac{2}{\hbar}\left(\frac{\mu}{k}\right)^{1/2} E \qquad (4.10)$$

is a unitless energy parameter.

Integration of Eq. (4.9) does not take into account the normalization condition. That additional constraint is included by introducing a second independent variable u equal to the normalization integral,

$$u = \int \psi^2\,dx'.$$

The differential equation for u is

$$\frac{du}{dx'} = \psi^2. \tag{4.11}$$

The program Schroed1 includes the variable u and offers the normalization option. Normalization is correct if u has a value of one when the
integration is complete.

Example 4-1. Use the program Schroed1 to confirm that the reduced
Schrödinger equation for the harmonic oscillator, Eq. (4.9), behaves as it
should for $E' = 1$, 3, 5, and 7, but not for any other values of E' in this
range.

Answer. The program Schroed1 uses the function NDSolve to integrate
Eqs. (4.9) and (4.11). The first two lines of code in NDSolve state the two
equations with x' and E' represented by x1 and E1, $\psi(x')$ by psi[x1],
$d\psi/dx'$ by psi'[x1], $d^2\psi/dx'^2$ by psi''[x1], and du/dx' by u'[x1].
The amplitude of the plotted wave function is controlled by changing the
initial value of $d\psi/dx'$, called slope0 in the program, as instructed in the
program's comment lines. Values of E' are calculated in the program with
$E' = 2v + 1$. Run the program with v given integer values and note that
$\psi(x')$ approaches small magnitudes as $x' \to \pm \infty$. Try some noninteger
values of v (even as slightly noninteger as 0.00001) and notice that $\psi(x')$
becomes very large in magnitude for large values of x'. Physically this describes an unstable molecule with a high probability for large atomic displacements, obviously not what we are looking for. The program demonstrates that wave functions calculated with Eq. (4.9) are "well behaved" only
if v in $E' = 2v + 1$ has *precisely* integer values and $E' = 1, 3, 5, 7 \ldots$. From
Eq. (4.10) conclude that

$$E = h\omega_e(v + 1/2),$$

with $v = 0, 1, 2, \ldots$, and $\omega_e = \dfrac{1}{2\pi}\left(\dfrac{k}{\mu}\right)^{1/2}$. This is the energy equation (4.5)
mentioned before without verification. Energies calculated with Eq. (4.5) are
energy eigenvalues and the corresponding well-behaved wave functions are
energy eigenfuncions.

For the Morse Anharmonic Oscillator

Here we consider the Schrödinger equation (4.4) with the Morse potential
energy function (4.3) substituted for $V(x')$. For the calculation the equation

is reduced to

$$\frac{d^2\psi}{dx'^2} = [V'(x') - E']\psi, \tag{4.12}$$

with x' and E' defined as before, and Eq. (4.3) simplified with

$$\beta = \sqrt{\frac{k}{2D_e}}, \tag{4.13}$$

so

$$V'(x') = D_e'(1 - e^{-x'/\sqrt{D_e'}})^2, \tag{4.14}$$

with

$$D_e' = \frac{2D_e}{\hbar}\left(\frac{\mu}{k}\right)^{1/2}.$$

The program Schroed2 integrates Eq. (4.12). Energy eigenvalues and eigenfunctions are obtained as they were for the harmonic oscillator in Example 4.1. The equation

$$E = \hbar\omega_e(v + 1/2) - x_e\hbar\omega_e(v + 1/2)^2, \tag{4.15}$$

with $x_e = \hbar\omega_e/4D_e = 1/2D_e'$, and v again given integer values, expresses energy eigenvalues. This can also be written

$$E' = (2v + 1) - (1/4D_e')(2v + 1)^2. \tag{4.16}$$

For an Electron in a Well

Now we integrate Eq. (4.4) with $V(x)$ expressed by Eqs. (4.7). In this case the equation reduces to

$$-\frac{d^2\psi}{dx'^2} + V'(x')\psi = E'\psi,$$

or

$$\frac{d^2\psi}{dx'^2} = [V'(x') - E']\psi, \tag{4.17}$$

where $x' = 2x/a$ and

$$V'(x') = \frac{\pi^2}{4}\frac{8m_ea^2}{h^2}V(x') \tag{4.18}$$

$$E' = \frac{\pi^2}{4}\frac{8m_ea^2}{h^2}E. \tag{4.19}$$

Half of the eigenvalues of the energy-related quantity E' are determined by

$$E' + E' \tan^2(\sqrt{E'}) = V_0', \qquad (4.20)$$

and the other half by

$$E' + \frac{E'}{\tan^2(\sqrt{E'})} = V_0', \qquad (4.21)$$

with

$$V_0' = \frac{\pi^2}{4} \frac{8 m_e a^2 V_0}{h^2}. \qquad (4.22)$$

Equations (4.20) and (4.21) are solved by numerical methods (see Exercise 4-4 and the program Well), and they have many roots, some of which are not significant in the eigenvalue problem. See Exercise 4-5 for an introduction to the program Schroed3, which integrates Eq. (4.17) and tests for valid energy eigenvalues.

4.3   Orbitals

The Schrödinger equation is the master equation in quantum chemistry, but mathematically speaking it has a serious flaw: it cannot be solved for most atomic and molecular problems without approximations. For many years the dominant approximation method used by quantum chemists has relied on the *orbital concept*.

An orbital is simply an atomic or molecular electronic state that can accommodate one electron—or two electrons if they have different spin states. Electronic structures of atoms and molecules are calculated with *atomic orbitals* and *molecular orbitals*. Shapes and sizes of atomic orbitals are found by solving approximate versions of the Schrödinger equation stated for atoms. Then molecular orbitals are formulated as linear combinations of atomic orbitals accessible to the atoms found in the molecule.

You can run the program Aorbital to display shapes and sizes of atomic orbitals for hydrogen atoms. Atomic orbitals for other atoms are similar in shape to these hydrogen atomic orbitals, but different in size. The program simulates data that might be obtained with the "gamma-ray microscope" Heisenberg invented for a thought experiment demonstrating one aspect of his uncertainty principle. In each "measurement" with the gamma-ray microscope, a spot is recorded where an electron is "located" by the microscope. The measurement drastically alters the atom, so the microscope must look at another atom of the same kind in the same state to record another spot where an electron is found. The program simulates the composite spot pattern that would be obtained if millions of atoms were put under the Heisenberg microscope. The pattern is bright where electrons are likely to be found and dark where they are unlikely to be found. In this way a *statistical* picture is obtained which defines the habitat of electrons occupying a certain kind of atomic orbital.

The program Morbital is similar except that is simulates sizes and shapes of σ and π molecular orbitals accessible to electrons in H_2^+ molecules. Similar patterns are found in larger diatomic molecules.

The program Duality demonstrates in another way the statistical basis of quantum theory. Duality simulates the interference-diffraction pattern obtained when a monoenergetic electron beam is passed through two closely-spaced slits. No evidence of interference-diffraction effects, therefore no evidence of wave behavior can be seen in observations of only a few electrons, but the pattern becomes evident if about 1000 electrons are observed in the beam. This makes the point that wave properties of electron beams can only be seen in the behavior of many electrons, and not at all in observations of a few electrons.

4.4   Molecular Mechanics

Molecules have four modes of motion that can cause changes in the molecule's energy. They are *translational motion*, in which the entire molecule moves from one place to another; *rotational motion*, in which the entire molecule rotates around its axes; *vibrational motion*, in which the molecule's nuclei move with respect to each other; and *electronic motion*, involving orbital and spin motion of the molecule's electrons. We should also mention *nuclear spin motion*. If no external fields are applied (for the moment we assume that there are none), changes in this mode do not have appreciable effects on the molecule's energy. Changes in molecular energy are now are chief concern, so we lose nothing important by omitting the nuclear spin mode from our list.

4.5   Energy Levels

Rotational

Energies for rotational levels of diatomic molecules are given by

$$E(J) = hc\tilde{B}_v J(J + 1) - hc\tilde{D}J^2(J + 1)^2, \tag{4.23}$$

where J is a rotational quantum number ($J = 0, 1, 2, \ldots$), $\tilde{B}_v$ is the *rotational constant* for the vth vibrational state, and $\tilde{D}$ is another constant. The term containing $\tilde{D}$ calculates the *centrifugal distortion* of a rotating diatomic molecule, an effect which is small enough so we can often ignore it. Both $\tilde{B}_v$ and $\tilde{D}$ are given in wavenumber units (cm^{-1}) units. (The overhead tilde, as in $\tilde{B}_v$, will always denote wavenumbers). The rotational constant $\tilde{B}_v$ depends on the vibrational quantum number v according to

$$\tilde{B}_v = \tilde{B}_e - \tilde{\alpha}_e(v + 1/2), \tag{4.24}$$

where $\tilde{\alpha}_e$ is a further empirical parameter and

$$\tilde{B}_e = \frac{h}{8\pi^2 I_e}, \tag{4.25}$$

with

$$I_e = \mu R_e^2. \tag{4.26}$$

This introduces the hypothetical "equilibrium" bond length R_e between atoms in a diatomic molecule that has halted its vibrational motion, and I_e is the equibrium moment of inertia.

Equation (4.23) calculates rotational energies of the levels "stacked" on the vth vibrational energy level. The vibrational term is omitted in Eq. (4.23), but it will be considered in the next subsection.

The program Ro1 calculates and plots rotational energy levels for diatomic molecules. Data are supplied to the program in wavenumber (cm^{-1}) units, as is customary in spectroscopy, but energy levels are calculated and plotted in electron volts. We will follow this practice in other programs that calculate and plot energy levels. Run the program and note the pattern of widening spaces between energy levels as the energy increases. Also notice that rotational energy levels are very closely spaced; they are plotted on an meV energy scale.

Diatomic molecules rotate around two axes that are perpendicular to the axis joining the atoms. Because practically all of the molecule's mass (in the nuclei) is located on the interatomic axis, rotation around that axis develops an insignificant amount of energy (not counting the energy resulting from electronic rotational motion, which is regarded as a contribution to the molecule's electronic energy). Thus, in effect, a diatomic molecule has only two rotational modes, and the moment of inertia for both modes is expressed by Eq. (4.26). A polyatomic molecule (any molecule not monatomic or diatomic) behaves the same way if it is linear, that is, if its nuclei are all located on the same axis. Nonlinear polyatomic molecules, on the other hand, always have three modes of rotational motion and three nonzero principal moments of inertia which are conventionally labeled I_A, I_B and I_C.

The rotational behavior of polyatomic (and diatomic) molecules is classified according to how the three principal moments of inertia compare:

1. *Linear molecules* are defined by $I_A = 0$, $I_B = I_C$, and the A axis coinciding with the interatomic axis (e.g., C_2H_2).
2. *Spherical-top molecules* are defined by $I_A = I_B = I_C$ (e.g., CH_4).
3. *Oblate symmetric-top molecules* are defined by $I_A = I_B < I_C$ and the C axis coinciding with the molecule's main rotation axis (e.g., C_6H_6).

4. *Prolate symmetric-top molecules* are defined by $I_A < I_B = I_C$ and the A axis coinciding with the molecule's main rotation axis (e.g., CH_3Cl).
5. *Asymmetric-top molecules* are defined by $I_A < I_B < I_C$ (e.g., H_2O).

The program Abc Calculates principal moments of inertia for molecules of any kind beginning with data on the coordinates of atoms in the molecule. Here is an example.

Example 4-2. Data are given below for atomic coordinates in $^{14}N^1H_3$. The origin is located at the center of the nitrogen atom. Use the program Abc to calculate principal moments of inertia for the molecule. What is the rotational classification of the shape of the molecule?

i	M_i (g mol^{-1})	$x_i/Å$	$y_i/Å$	$z_i/Å$
1	14.003	0	0	0
2	1.0078	0	0.940	−0.381
3	1.0078	0.814	−0.470	−0.381
4	1.0078	−0.814	−0.470	−0.381

Answer. Data entries in Abc, taken from the above table and beginning on the third line of code, are

```
Data = {{14.003, 0, 0, 0},
        {1.0078, 0, .940, -.381},
        {1.0078, .814, -.470, -.381},
        {1.0078, -.814, -.470, -.381}};
```

The program calculates $I_A = I_B = 2.817 \times 10^{-47}$ kg m^2 and $I_C = 4.436 \times 10^{-47}$ kg m^2. Because $I_A = I_B < I_C$, the rotating molecule has an oblate symmetric-top shape.

Energy levels for nonlinear polyatomic molecules are calculated with two quantum numbers, J and K. Our examples consider symmetric-top molecules whose energy equations are:

for a prolate symmetric top

$$E(J, K) = hc\tilde{B}J(J+1) + hc(\tilde{A} - \tilde{B})K^2, \tag{4.27}$$

and for an oblate symmetric top

$$E(J, K) = hc\tilde{B}J(J+1) + hc(\tilde{C} - \tilde{B})K^2. \tag{4.28}$$

In these equations, $\tilde{A}$, $\tilde{B}$, and $\tilde{C}$ are rotational constants defined by

$$\tilde{A} = \frac{h}{8\pi^2 I_A c}$$

$$\tilde{B} = \frac{h}{8\pi^2 I_B c}$$

$$\tilde{C} = \frac{h}{8\pi^2 I_C c}.$$

$\tilde{A}$, $\tilde{B}$, and $\tilde{C}$ are all given in wavenumber units (cm^{-1}), and $\tilde{C} \leq \tilde{B} \leq \tilde{A}$ since $I_A \leq I_B \leq I_C$. The moments of inertia I_A, I_B, and I_C (and the rotational constants $\tilde{A}$, $\tilde{B}$, and $\tilde{C}$) depend slightly on the vibrational quantum number v, but we will not include that detail. We also ignore the effects of Coriolis coupling.

The program Ro2 plots energy levels for oblate and prolate symmetric-top molecules. Run the program for both cases and note differences in the energy level patterns. How do Eqs. (4.27) and (4.28) account for these differences?

Rotational–Vibrational

We now elaborate the energy calculations by including the vibrational energy to which the rotational energies just calculated are added. Neglecting the centrifugal-distortion contribution [the term containing D in Eq. (4.23)], the rotational–vibrational energy equation for a diatomic molecule is

$$E(J, v) = hc\tilde{\omega}_e(v + 1/2) - hc\tilde{\omega}_e x_e(v + 1/2)^2 + hc\tilde{B}_v J(J + 1), \qquad (4.29)$$

in which

$$\tilde{\omega}_e = \frac{1}{2\pi c}\left(\frac{k}{\mu}\right)^{1/2}$$

[Eq. (4.6)], x_e is a small unitless constant and $\tilde{B}_v$ is again given by Eq. (4.24).

The program Rovi1 plots energy levels calculated with Eq. (4.29) for two vibrational levels. Run the program and notice the energy scale and the pattern of rotational levels stacked on vibrational levels. The hierarchical arrangements of energy levels is beginning to be apparent: we see rotational levels stacked on vibrational levels, and in the next subsection vibrational levels stacked on electronic levels.

Vibrational–Electronic

Consider two states of electronic motion for a diatomic molecule whose energies without rotational and vibrational contributions are E_e'' and E_e' for the lower and upper states, or in wavenumber units, $\tilde{T}_e'' = E_e''/hc$ and $\tilde{T}_e' = E_e'/hc$. Including vibrational contributions we have for the lower and upper energies,

$$E(v'', \Lambda'') = hc\tilde{T}_e'' + hc\tilde{\omega}_e''(v'' + 1/2) - hc\tilde{\omega}_e''x_e''(v'' + 1/2)^2 \quad (4.30)$$

$$E(v', \Lambda') = hc\tilde{T}_e' + hc\tilde{\omega}_e'(v' + 1/2) - hc\tilde{\omega}_e'x_e'(v' + 1/2)^2, \quad (4.31)$$

where v'', Λ'' and v', Λ' are vibrational and electronic quantum numbers for the lower and upper states.

The program Vie11 plots vibrational–electronic, or *vibronic*, energy levels according to these equations. Run the program and note the pattern of the spacings. Notice also that energies plotted now cover the range 0–10 eV, compared to the previous plots covering the range 0–200 meV (rotational levels) and 0–1 eV (vibrational levels).

The program Morse also plots vibronic energy levels for diatomic molecules and adds to the physical picture by superimposing the levels on plots of the Morse equation (4.3) for the lower and upper electronic states. To avoid congestion, only one third of the levels are plotted. The line for each level extends between the classical *turning points*, that is, the points where vibrational motion turns from stretching to compressing, and also where (in classical theory) the kinetic energy of the motion is zero and the potential energy has its maximum value. The Morse equation includes an energy factor D_e, which measures the dissociation energy from the minimum value of the potential energy $V(x)$. Morse uses another dissociation energy D_0 which measures from the energy level of the molecule's zero-point vibrational state. The two dissociation energies are related by

$$D_e = D_0 + h\omega_e/2 - h\omega_e x_e/4. \quad (4.32)$$

Rotational–Vibrational–Electronic

We complete this picture of the molecular hierarchy of energy levels by writing energy equations for lower and upper *rovibronic* states located by the rotational, vibrational, and electronic quantum numbers J'', v'', Λ'' for the lower state and J', v', Λ' for the upper state. These equations are composites of Eqs. (4.29) to (4.31),

$$E(J'', v'', \Lambda'') = hc\tilde{T}_e'' + hc\tilde{\omega}_e''(v'' + 1/2) - hc\tilde{\omega}_e''x_e''(v'' + 1/2)^2$$
$$+ hc\tilde{B}_v''J''(J'' + 1) \quad (4.33)$$

$$E(J', v', \Lambda') = hc\tilde{T}_e' + hc\tilde{\omega}_e'(v' + 1/2) - hc\tilde{\omega}_e'\tilde{x}_e'(v' + 1/2)^2$$
$$+ hc\tilde{B}_v'J'(J' + 1). \quad (4.34)$$

4.6   The Hartree–Fock Equation

The Schrödinger equation has two fundamental limitations: it ignores the principles of relativity and does not recognize the existence of electron spin. The first omission is usually not important in physical chemistry because atomic and molecular interaction energies are low enough to make relativistic corrections negligible. The electron spin concept, on the other hand, is profoundly important in a major branch of physical chemistry, the theory of molecular structure. Practitioners of quantum chemistry, who focus on molecular calculations, demand broader equations which incorporate the effects of electron spin.

The broader theory begins with the orbital approximation. Each of a molecule's electrons is assigned to an orbital, and that permits calculation of the electron's kinetic energy, attraction potential energy of interaction with the molecule's nuclei, and repulsion potential energy of interaction with electrons in other orbitals. When two orbitals have spatial regions in common, the theory corrects the interaction energy to conform with the Pauli requirement that electrons in the same spin state cannot occupy the same region. The equation that does this calculation was developed by Hartree and Fock in the 1920s and 1930s. Its mathematical form is

$$\hat{F}\phi_i = E\phi_i, \tag{4.35}$$

where ϕ_i is a molecular orbital formulated as a linear combination of atomic orbitals u_a,

$$\phi_i = \sum_a c_{ai} u_a, \tag{4.36}$$

E is a molecular orbital energy and $\hat{F}$, called a *Fock operator*, stands for the many differential and algebraic operators that represent molecular kinetic and potential energy. The *Hartree–Fock equation* (4.35) repairs some of the limitations inherent in the Schrödinger equation, but it has limitations of its own connected with the orbital approximation. The further theory is beyond our scope, however. Our quantum chemistry story begins and ends with a brief account of matrix versions of the Hartree–Fock equation.

4.7   Matrix Equations

For the sake of efficiency in computer calculations, the Hartree–Fock equation is usually manipulated into a matrix format. Three matrices are important: the *Fock matrix* $\mathbf{F}$, *orbital matrix* $\mathbf{C}$, and *overlap matrix* $\mathbf{S}$. If molecular orbitals are constructed from three atomic orbitals u_1, u_2, and u_3, the Fock matrix is

$$\mathbf{F} = \begin{pmatrix} F_{11} & F_{12} & F_{13} \\ F_{12} & F_{22} & F_{23} \\ F_{13} & F_{23} & F_{33} \end{pmatrix}, \tag{4.37}$$

in which the matrix elements are integrals. For example,

$$F_{11} = \int u_1 \hat{F} u_1 \, d\tau \quad \text{and} \quad F_{12} = \int u_1 \hat{F} u_2 \, d\tau,$$

with integration covering the space of the molecule. The corresponding overlap matrix is

$$\mathbf{S} = \begin{pmatrix} S_{11} & S_{12} & S_{13} \\ S_{12} & S_{22} & S_{23} \\ S_{13} & S_{23} & S_{33} \end{pmatrix}, \tag{4.38}$$

where, for example,

$$S_{11} = \int u_1 u_1 \, d\tau \quad \text{and} \quad S_{12} = \int u_1 u_2 \, d\tau.$$

When an orbital calculation is done with the 3×3 Fock matrix (4.37) three molecular orbitals are obtained whose coefficients in Eq. (4.36) are expressed as columns in the orbital matrix $\mathbf{C}$,

$$\mathbf{C} = \begin{pmatrix} c_{11} & c_{12} & c_{13} \\ c_{21} & c_{22} & c_{23} \\ c_{31} & c_{32} & c_{33} \end{pmatrix}. \tag{4.39}$$

The matrix version of the Hartree–Fock equation is simply

$$\mathbf{FC} = \mathbf{SCE}, \tag{4.40}$$

in which $\mathbf{E}$ is a diagonal matrix with the orbital energy eigenvalues $E_1, E_2, \ldots$ located on its diagonal. If the Fock matrix is (4.37), $\mathbf{E}$ is

$$\mathbf{E} = \begin{pmatrix} E_1 & 0 & 0 \\ 0 & E_2 & 0 \\ 0 & 0 & E_3 \end{pmatrix}. \tag{4.41}$$

The strategy of a molecular orbital calculation is to begin with a suitable Fock matrix $\mathbf{F}$, then to calculate the energy matrix $\mathbf{E}$ and simultaneously the orbital matrix $\mathbf{C}$. That is much easier said than done, however. Each Fock matrix element F_{ab} is complicated because it evaluates electrostatic and spin-related interactions between an orbital overlap $u_a u_b$ and all the other overlaps in the molecule. Multitudes of integrals, some of them very complicated, have to be determined to calculate these interactions. To make matters worse, the Fock matrix is unknown at the beginning of the calculation because it depends on the orbital matrix $\mathbf{C}$, which is also unknown at the outset. The calculation is saved by first approximating the Fock matrix $\mathbf{F}$ as the *core matrix* $\mathbf{H}$, which neglects the most difficult part of the calculation, that concerned with interelectron electrostatic interactions. This permits calculation of an approximate orbital matrix $\mathbf{C}$, which is used to obtain a better Fock matrix, and so on through a series of iterative steps until a result with the desired accuracy is obtained.

One molecular orbital method, introduced in 1930 by Hückel, avoids all of these complications by resorting to some drastic simplifications. Hückel's method concerns π electrons found in molecules containing conjugated carbon chains (e.g., butadiene). It assumes that the overlap matrix S can be expressed as the identity matrix I, so Eq. (4.40) becomes

$$FC = CE, \tag{4.42}$$

and that the Fock matrix has only two kinds of nonzero elements, α and β, which do not require iterations for their calculation. The quantity α is the energy an electron would have if it were confined to a carbon $2p\pi$ carbon atomic orbital, and β assesses the $2p\pi$ bond energy developed between two adjacent carbon atoms in the chain. The Hückel version of the Fock matrix for a three-carbon conjugated chain (the allyl free radical) is

$$F = \begin{pmatrix} \alpha & \beta & 0 \\ \beta & \alpha & \beta \\ 0 & \beta & \alpha \end{pmatrix}. \tag{4.43}$$

The rules are to place α's on the diagonal of the Fock matrix, β's in positions corresponding to adjacent carbons in the chain, and 0's elsewhere.

To solve Eq. (4.42) in the Hückel procedure multiply on the left by $\tilde{C}$, the transpose of C,

$$\tilde{C}FC = \tilde{C}CE,$$

and since $\tilde{C}C = I$, obtain

$$\tilde{C}FC = E. \tag{4.44}$$

This tells us that the orbital matrix diagonalizes the Fock matrix F in the transformation $\tilde{C}FC$ and places the energy eigenvalues on the diagonal. The calculation begins with the Fock matrix expressed, for example, as in Eq. (4.43), and the strategy is to find a matrix that accomplishes the diagonalization. That matrix is the orbital matrix and the diagonal elements in the diagonalized matrix are the energy eigenvalues. The program Hueckel carries out this procedure on Fock matrices which are further reduced as described in Exercise 4-21.

The full Hartree–Fock procedure requires solution of Eq. (4.40), including the overlap matrix S. This more complicated matrix equation does not permit us to proceed by diagonalizing the Fock matrix F. But a few simple matrix manipulations put the equation into a form that allows use of standard diagonalization procedures.

Introduce the square-root matrix $S^{1/2}$ (this matrix multiplied by itself equals S) and multiply both sides of Eq. (4.40) on the left by $S^{-1/2}$, the inverse of $S^{1/2}$,

$$S^{-1/2}FC = S^{-1/2}SCE.$$

A factor of $\mathbf{I}$, the identity matrix, can be introduced anywhere in this equation without changing it. Write

$$\mathbf{S}^{-1/2}\mathbf{FIC} = \mathbf{S}^{-1/2}\mathbf{ISCE},$$

note that $\mathbf{I} = \mathbf{S}^{1/2}\mathbf{S}^{-1/2}$, that $\mathbf{S} = \mathbf{S}^{1/2}\mathbf{S}^{1/2}$, and write

$$\mathbf{S}^{-1/2}\mathbf{FS}^{-1/2}\mathbf{S}^{1/2}\mathbf{C} = \mathbf{S}^{-1/2}\mathbf{S}^{1/2}\mathbf{S}^{-1/2}\mathbf{S}^{1/2}\mathbf{S}^{1/2}\mathbf{CE}$$

$$= \mathbf{S}^{1/2}\mathbf{CE}.$$

Now define a transformed orbital matrix $\mathbf{C}'$,

$$\mathbf{C}' = \mathbf{S}^{1/2}\mathbf{C}, \tag{4.45}$$

a transformed Fock matrix $\mathbf{F}'$,

$$\mathbf{F}' = \mathbf{S}^{-1/2}\mathbf{FS}^{-1/2}, \tag{4.46}$$

and put the Fock matrix equation in the same form as Eq. (4.42),

$$\mathbf{F}'\mathbf{C}' = \mathbf{C}'\mathbf{E}. \tag{4.47}$$

The further calculation proceeds just as it does in Hückel theory except that the transformed Fock matrix $\mathbf{F}'$ is diagonalized and the calculation produces the transformed orbital matrix $\mathbf{C}'$. To obtain the desired orbital matrix $\mathbf{C}$ multiply Eq. (4.45) on the left by $\mathbf{S}^{-1/2}$,

$$\mathbf{S}^{-1/2}\mathbf{C}' = \mathbf{S}^{-1/2}\mathbf{S}^{1/2}\mathbf{C},$$

or

$$\mathbf{C} = \mathbf{S}^{-1/2}\mathbf{C}'. \tag{4.48}$$

The matrix $\mathbf{E}$ containing energy eigenvalues does not require transformation.

The computer program Hartree implements a full Hartree–Fock calculation for the diatomic molecule LiH. It reads from the file Chap4.m a large collection of preliminary integrals, calculates the $\mathbf{S}^{1/2}$ and $\mathbf{S}^{-1/2}$ matrices, and then does its further work in iterative steps. Each step begins with an approximate Fock matrix $\mathbf{F}$, then transforms $\mathbf{F}$ to $\mathbf{F}'$, diagonalizes $\mathbf{F}'$, thus calculating $\mathbf{C}'$ and $\mathbf{E}$, transforms $\mathbf{C}'$ to $\mathbf{C}$, recalculates $\mathbf{F}$, goes through another step, and continues in this way through a prescribed number of iterations.

4.8   Exercises

Schrödinger Equations Solved

4-1 The program Schroed1 obtains vibrational eigenfunctions by numerical integration of the Schrödinger equation for the harmonic oscillator. The same results can also be obtained by solving the equation analytically. The first two eigenfunctions obtained this way (for $v = 0$ and 1) are

$$\psi_0 = (1/\pi^{1/4})e^{-x'^2/2}$$

$$\psi_1 = -(2^{1/2}/\pi^{1/4})x'e^{-x'^2/2}.$$

Write a program that plots these functions and compares with results obtained by numerical integration with the program Schroed1.

4-2 Run the program Schroed1 for $v = 0, 1, 2$, and 3. Note occurrences of *nodes*, where the wave function crosses the x axis and has zero values. For a given value of v how many nodes does the wave function have?

4-3 Run the program Schroed2, which integrates Eq. (4.12) for the Morse anharmonic oscillator, with $v = 0$, 1, 2, 3, and $D'_e = 25$. Note differences between wave functions obtained with this equation and those obtained with Eq. (4.9) for the harmonic oscillator.

4-4 Energy eigenvalues for the electron-in-the-well problem are calculated with Eqs. (4.20) and (4.21). The parameter V'_0 in those equations we now write $V'_0 = \alpha \pi^2/4$, where the unitless parameter α is defined $\alpha = 8m_e a^2 V_0/h^2$, with a and V_0 the width and depth of the well. The program Well numerically calculates roots of Eqs. (4.20) and (4.21) for a given value of α. Run the program for $\alpha = 100$ and use these results in the next exercise.

4-5 Run the program Schroed3 to test the list of roots obtained in the last exercise concerning whether or not they are energy eigenvalues. Try each root in turn until you have found four valid energy eigenvalues. Do not bother with the normalization.

4-6 Use the program Abc and the atomic coordinates given below for $^{12}C^1H_4$ to calculate the molecule's principal moments of inertia. What is the rotational classification of the molecule's shape?

i	M_i (g mol^{-1})	$x_i/\text{Å}$	$y_i/\text{Å}$	$z_i/\text{Å}$
1	12.000	0	0	0
2	1.0078	0.6291	0.6291	0.6291
3	1.0078	−0.6291	0.6291	−0.6291
4	1.0078	0.6291	−0.6291	−0.6291
5	1.0078	−0.6291	−0.6291	0.6291

4-7 Use the program Abc and atomic coordinates given below for $^{12}C^1H_3{}^{35}Cl$ to calculate the molecule's principal moments of inertia. What is the rotational classification of the molecule's shape?

i	M_i (g mol^{-1})	x_i/Å	y_i/Å	z_i/Å
1	12.000	−0.1792	−0.7096	0
2	34.969	−0.1792	1.0547	0
3	1.0078	0.8493	−1.0825	0
4	1.0078	−0.6935	−1.0825	0.8908
5	1.0078	−0.6935	−1.0825	−0.8908

4-8   Use the program Abc and the atomic coordinates given below for $^{12}C_2{}^{1}H_2$ to calculate the molecule's principal moments of inertia. What is the rotational classification of the molecule's shape?

i	M_i (g mol^{-1})	x_i/Å	z_i/Å	z_i/Å
1	12.000	0	0	0
2	12.000	1.208	0	0
3	1.0078	−1.058	0	0
4	1.0078	2.266	0	0

4-9   Use the program Ro1 to calculate and plot rotational energy levels for $Cl_2(g)$ in its ground vibrational and electronic state. Rotational and vibrational constants for $Cl_2(g)$ are $\tilde{B}_e = 0.2439$ cm^{-1} and $\tilde{\alpha}_e = 0.00149$ cm^{-1}.

4-10   Use the program Ro1 to calculate and plot rotational energy levels for $N_2(g)$ in its ground vibrational and electronic state. Rotational and vibrational constants for $N_2(g)$ are $\tilde{B}_e = 1.99824$ cm^{-1} and $\tilde{\alpha}_e = 0.017318$ cm^{-1}.

4-11   Use the program Ro1 to calculate and plot rotational energy levels for $I_2(g)$ in its ground vibrational and electronic state. Rotational and vibrational constants for $I_2(g)$ are $\tilde{B}_e = 0.03737$ cm^{-1} and $\tilde{\alpha}_e = 0.000114$ cm^{-1}.

4-12   Use the program Ro2 to calculate and plot rotational energy levels for $NF_3(g)$ in its ground vibrational and electronic state. Principal moments of inertia for $NF_3(g)$ are $I_A = I_B = 7.859 \times 10^{-46}$ kg m^2 and $I_C = 1.437 \times 10^{-45}$ kg m^2.

4-13   Use the program Ro2 to calculate and plot rotational energy levels for $CCIF_3(g)$ in its ground vibrational and electronic state. Principal moments of inertia for $CCIF_3(g)$ are $I_A = 1.467 \times 10^{-45}$ kg m^2, $I_B = I_C = 2.520 \times 10^{-45}$ kg m^2.

4-14   Use the program Rovil to calculate and plot rotational–vibrational energy levels for $^{35}Cl_2(g)$ in its ground electronic state and in vibrational states corresponding to $v = 0$ and $v = 1$. Rotational and vibrational constants are

$$\tilde{\omega}_e = 559.72 \text{ cm}^{-1}$$

$$\tilde{\omega}_e x_e = 2.675 \text{ cm}^{-1}$$

$$\tilde{\alpha}_e = 0.00149 \text{ cm}^{-1}$$

$$\tilde{B}_e = 0.2439 \text{ cm}^{-1}.$$

4-15 Use the program Rovil to calculate and plot rotational–vibrational energy levels for $^{14}N_2(g)$ in its ground electronic state and in vibrational states corresponding to $v = 0$ and $v = 1$. Rotational and vibrational constants are

$$\tilde{\omega}_e = 2358.57 \text{ cm}^{-1}$$

$$\tilde{\omega}_e x_e = 14.324 \text{ cm}^{-1}$$

$$\tilde{\alpha}_e = 0.017318 \text{ cm}^{-1}$$

$$\tilde{B}_e = 1.99824 \text{ cm}^{-1}.$$

4-16 Use the program Rovil to calculate and plot rotational–vibrational energy levels for $^{127}I_2(g)$ in its ground electronic state and vibrational states corresponding to $v = 0$ and $v = 1$. Rotational and vibrational constants are

$$\tilde{\omega}_e = 214.50 \text{ cm}^{-1}$$

$$\tilde{\omega}_e x_e = 0.615 \text{ cm}^{-1}$$

$$\tilde{\alpha}_e = 0.000114 \text{ cm}^{-1}$$

$$\tilde{B}_e = 0.03737 \text{ cm}^{-1}.$$

4-17 Use the program Viell to calculate and plot vibrational–electronic (vibronic) energy levels for two $^{12}C^{16}O(g)$ electronic states, $X^1\Sigma^+$ and $A^1\Pi$. Vibrational and electronic constants are:

$$\tilde{\omega}_e'' = 2169.8136 \text{ cm}^{-1}$$

$$\tilde{\omega}_e'' x_e'' = 13.28831 \text{ cm}^{-1}$$

$$\tilde{T}_e'' = 0$$

$$\tilde{\omega}_e' = 1518.2 \text{ cm}^{-1}$$

$$\tilde{\omega}_e' x_e' = 19.40 \text{ cm}^{-1}$$

$$\tilde{T}_e' = 65075.8 \text{ cm}^{-1}.$$

4-18 Use the program Viell to calculate and plot vibrational–electronic (vibronic) energy levels for two $^{14}N_2(g)$ electronic states, $A^3\Sigma_u^+$ and $B^3\Pi_g$. Vibrational and electronic constants are

$$\tilde{\omega}_e'' = 1460.64 \text{ cm}^{-1}$$

$$\tilde{\omega}_e'' x_e'' = 13.872 \text{ cm}^{-1}$$

$$\tilde{T}_e'' = 50203.6 \text{ cm}^{-1}$$

$$\tilde{\omega}_e' = 1733.39 \text{ cm}^{-1}$$

$$\tilde{\omega}_e' x_e' = 14.122 \text{ cm}^{-1}$$

$$\tilde{T}_e' = 59619.4 \text{ cm}^{-1}.$$

4-19 Use the program Morse to calculate and plot Morse curves and vibrational energy levels for two $^{12}C^{16}O(g)$ electronic states, $X^1\Sigma^+$ and $A^1\Pi$. Vibrational and electronic constants for these states are given in Exercise 4-17. Note also that $D_e'' = 9.05$ eV and $D_e' = 4.34$ eV, and that $R_e'' = 0.1128323$ nm and $R_e' = 0.12353$ nm.

4-20 Use the program Morse to calculate and plot Morse curves and vibrational energy levels for two $^{14}N_2(g)$ electronic states, $A^3\Sigma_u^+$ and $B^3\Pi_g$. Vibrational and electronic constants for these states are given in Exercise 4-18. Note also that $D_e'' = 3.6$ eV and $D_e' = 4.0$ eV, and that $R_e'' = 0.12866$ nm and $R_e' = 0.12126$ nm.

Matrix Equations

4-21 The Hückel calculation of molecular orbitals can be expressed $\mathbf{FC} = \mathbf{CE}$ [see Eq. (4.42)]. For the Hückel calculation of π molecular orbitals in the allyl radical the Fock matrix $\mathbf{F}$ is

$$\mathbf{F} = \begin{pmatrix} \alpha & \beta & 0 \\ \beta & \alpha & \beta \\ 0 & \beta & \alpha \end{pmatrix}.$$

This expression for $\mathbf{F}$ is unsatisfactory, however, because the Hückel procedue provides no evaluation of α and β. This problem is avoided by revising the original matrix equation so it has the form

$$\mathbf{F'C} = \mathbf{Cx},$$

in which

$$\mathbf{F} = \begin{pmatrix} 0 & -1 & 0 \\ -1 & 0 & -1 \\ 0 & -1 & 0 \end{pmatrix}$$

for the allyl calculation and

$$\mathbf{x} = \begin{pmatrix} x_1 & 0 & 0 \\ 0 & x_2 & 0 \\ 0 & 0 & x_3 \end{pmatrix},$$

where the x's are eigenvalues of the energy-related variable

$$x = (\alpha - E)/\beta.$$

The general recipe for formulating the Fock matrix in this scheme is to re-place the α's with 0's and the β's with -1's. When *this* matrix is diagonalized the diagonal elements are eigenvalues of x and $\mathbf{C}$ is again the orbital matrix. The program Hueckel does the Hückel (or Hueckel) calculation this way. Rows of the reduced matrix $\mathbf{F}'$ are entered in Hueckel, the program diag-onalizes $\mathbf{F}'$, and prints eigenvalues of x and corresponding columns of the orbital matrix $\mathbf{C}$ (i.e., the orbital coefficients). Use Hueckel to calculate the allyl π orbitals.

4-22   Use the Hueckel program to calculate π molecular orbitals for butadiene, a four-carbon conjugated straight chain hydrocarbon.

4-23   Use the Hartree program to make Hartree–Fock calculations for the molecule LiH containing 2, 4 and 6 electrons (i.e., for LiH^{2+}, LiH and LiH^{2-}) assuming (unrealistically) that the size of the molecules does not change when electrons are added to or subtracted from the neutral mole-cule LiH. Note the calculated total energy in each case and account for the differences.

5
Spectroscopy

The proof of quantum theory lies mainly in the data of spectroscopy. In-dividually the peaks of a molecule's spectrum represent emission or absorption transitions between energy levels, and collectively they build a richly detailed picture of the manifold of energy levels accessible to the molecule while it is engaged in one or more of its modes of motion. The effect of a photon absorbed by a molecule from a spectrometer beam depends on the energy of the photon. Microwave photons excite rotational transitions, infrared photons rotational and vibrational transitions, and visible or ultraviolet photons rotational, vibrational, and electronic transitions. Photons from a radio frequency source excite spin transitions in an applied magnetic field. Rotational spectra are illustrated in Sec. 5.1 (the program Ro3), rotational-vibrational spectra in Sec. 5.2 (programs Peaks, Rovi2, Symtop1, and Symtop2), vibrational-electronic spectra in Sec. 5.3 (the program Viel2), and rotational-vibrational-electronic spectra in Sec. 5.4 (the program Roviel). Spin resonance spectra are illustrated in Sec. 5.5 (programs 2dnmr, Nmr, and Esr). The chapter closes with an account of Fourier-transform techniques for analyzing spectrometer data (programs Irft and Nmrft).

5.1   Rotational Spectroscopy

Rotational energy levels for diatomic molecules are calculated with Eq. (4.23), which can be simplified without losing much accuracy by omitting the centrifugal distortion term,

$$E(J) = hc\tilde{B}_v J(J+1),\tag{5.1}$$

in which J is the rotational quantum number and the rotational constant $\tilde{B}_v$ is calculated with Eq. (4.24). A rotational transition $J_1 \leftarrow J_0$ between a lower rotational state located by J_0 and an upper state by J_1 is induced when the molecule absorbs a photon of energy $hv = E(J_1) - E(J_0)$, frequency

$v = [E(J_1) - E(J_0)]/h$, and wavenumber $\tilde{v} = [E(J_1) - E(J_0)]/hc$. Rotational spectra for diatomic molecules are restricted by the selection rule

$$\Delta J = J_1 - J_0 = +1.$$

The program Ro3 calculates and plots rotational spectra for diatomic molecules. It also supplies a direct interpretation of the spectrum by plotting horizontal lines for the rotational energy levels involved and vertical lines for the transitions dictated by the above selection rule.

5.2 Rotational–Vibrational Spectroscopy

For Diatomic Molecules

Rotational–vibrational energy levels for diatomic molecules are calculated with good accuracy by Eq. (4.29). A rotational-vibrational transition $J_1, 1 \leftarrow J_0, 0$ between the ground and first excited vibrational state (specified by $v = 0$ and 1) is induced by a photon of wavenumber

$$\tilde{v} = \frac{E(J_1, 1) - E(J_0, 0)}{hc}. \tag{5.2}$$

Rotational–vibrational transitions for diatomic molecules are restricted by the selection rules

$$\Delta J = J_1 - J_0 = -1 \text{ or } +1.$$

These two conditions account for the two branches, called P and R, found in rotational–vibrational spectra of diatomic molecules. The P *branch*, on the low-wavenumber side of the center of the spectrum, is defined by $\Delta J = -1$, and the R *branch*, on the high-wavenumber side, by $\Delta J = +1$. If $\Delta J = 0$ were allowed, a third branch would be seen near the center of the spectrum and would be called a Q *branch*. The center, calculated with $J_0 = J_1 = 0$ in Eqs. (4.29) and (5.2), is located at the wavenumber

$$\tilde{v}_{00} = \tilde{\omega}_e - 2\tilde{\omega}_e x_e. \tag{5.3}$$

The program Rovi2 calculates and plots rotational–vibrational spectra for diatomic molecules, and it interprets the spectra by plotting horizontal lines for the rotational-vibrational levels and vertical lines for the transitions.

For Polyatomic Molecules

Rotational–vibrational spectra for polyatomic molecules are considerably more complicated than those for diatomic molecules. One source of complexity is the requirement of more than one quantum number in the calculation of rotational energy levels for nonlinear molecules. We saw that in Sec. 4.5 and introduced Eqs. (4.27) and (4.28), which calculate rotational energy levels for prolate and oblate symmetric-top molecules. Another complication is that polyatomic molecules have two kinds of vibrational modes, characterized as *parallel* and *perpendicular*, in which the molecule's dipole moment changes are parallel or perpendicular to the main rotation axis.

For a symmetric-top molecule fundamental absorption transitions are represented by $J_1, K_1, 1 \leftarrow J_0, K_0, 0$, with $v = 0$ in the lower state and $v = 1$ in the upper. Corresponding wavenumbers for peaks in the spectrum are calculated with

$$\tilde{v} = \frac{E(J_1, K_1, 1) - E(J_0, K_0, 0)}{hc}. \tag{5.4}$$

We have seen that the rotational–vibrational spectrum of a diatomic molecule consists of bands of peaks separated into P and R branches interpreted by assigning $\Delta J = -1$ to the P branch and $\Delta J = +1$ to the R branch. Symmetric-top molecules display another level of complexity. Each band is a composite of a series of *sub-bands* which resemble entire bands in the spectra of diatomic molecules, with the difference that, in addition to P and R branches, most sub-bands have intense Q branches.

Sub-bands are interpreted by labeling them with a value of K_0 and P, Q, and R for $\Delta K = -1$, 0, and $+1$. Branches of the sub-bands are labeled as before: P, Q, and R for $\Delta J = -1$, 0, and $+1$. All of this is condensed into a convenient notation: ${}^P Q_5$ denotes the Q branch ($\Delta J = 0$) of the P sub-band ($\Delta K = -1$) for which $K_0 = 5$, and ${}^Q P_0$ the P branch ($\Delta J = -1$) of the Q sub-band ($\Delta K = 0$) for which $K_0 = 0$.

Selection rules for the two rotational quantum numbers J and K are: for parallel vibrational modes,

$$\Delta K = 0 \quad \text{and} \quad \Delta J = \pm 1 \quad \text{if } K_0 = 0$$

$$\Delta K = 0 \quad \text{and} \quad \Delta J = 0, \pm 1 \quad \text{if } K_0 > 0,$$

and for perpendicular vibrational modes,

$$\Delta K = +1 \quad \text{and} \quad \Delta J = 0, \pm 1 \quad \text{if } K_0 = 0$$

$$\Delta K = \pm 1 \quad \text{and} \quad \Delta J = 0, \pm 1 \quad \text{if } K_0 > 0.$$

The program Symtop1 calculates and plots a complete *parallel band* (i.e., generated by a parallel vibrational mode) for a prolate symmetric-top molecule. Energy levels are calculated with Eq. (4.27) and spectral wavenumbers with Eq. (5.4). The two rotational constants $\tilde{A}$ and $\tilde{B}$ are assumed to have slightly different values in the lower and upper vibrational states. Run the program and note the overall pattern of the spectrum. Then follow instructions in the program's comments and plot a few of the sub-bands separately, each labeled with $\Delta K = 0$ and a value of K_0. Note also the P, Q, and R branch structure of the individual sub-bands.

Another program, Symtop2, calculates and plots a *perpendicular band* (generated by a perpendicular vibrational mode) for a prolate symmetric-top molecule. Energy levels and wavenumbers are again calculated with Eqs. (4.27) and (5.4), but differences in the selection rules lead to a still more complicated spectrum. Run the program for the entire spectrum and then for individual sub-bands defined by $\Delta K = -1$ or $+1$ and values of K_0.

Data Analysis

Once you have observed a molecular spectrum, the next step is to extract from it information concerning the molecule's physical properties, such as its bond lengths and force constants. We include an example in which data from the rotational–vibrational spectrum of HCl(g) are analyzed. These wavenumbers fit Eqs. (4.29) and (5.2) reduced to

$$\tilde{\nu} = \tilde{\nu}_{00} + (\tilde{B}_1 + \tilde{B}_0)m + (\tilde{B}_1 - \tilde{B}_0)m^2. \tag{5.5}$$

In this equation $\tilde{B}_0$ and $\tilde{B}_1$ are given by Eq. (4.24), $\tilde{\nu}_{00}$ locates the center of the spectrum and m is a running index which labels peaks of the P branch with $m = -1, -2, \ldots$, peaks of the R branch with $m = 1, 2, \ldots$, and the value $m = 0$ is not allowed. The peaks next to the gap at the center of the spectrum are assigned $m = -1$ in the P branch and $m = 1$ in the R branch. Other values of m are assigned to the peaks consecutively. The next example shows how this is done.

Example 5-1. The program Peaks reads a file of observed wavenumbers for the rotational–vibrational spectrum of HCl(g), plots the data and calculates wavenumbers for all of the peak maxima. Run the program and notice first that the plot is a composite of two spectra, one for $^1H^{35}Cl$ and another for $^1H^{37}Cl$. The lighter chlorine isotope ^{35}Cl dominates, so the peaks of the $^1H^{35}Cl$ spectrum have higher intensities than those for $^1H^{37}Cl$, and they occur at slightly higher wavenumbers. Run Peaks, locate the $^1H^{35}Cl$ peaks, and assign each one a value of the index m in Eq. (5.5).

Answer. The program Peaks prepares a list of wavenumbers for peak maxima in the spectra for both $^1H^{35}Cl$ and $^1H^{37}Cl$. Distinguish the $^1H^{35}Cl$ peaks by their higher intensities and wavenumbers compared to the corresponding $^1H^{37}Cl$ peaks. The plot of the spectrum clearly shows the gap at the center of the spectrum. Assign $m = -1$ to 2865.0 cm^{-1} in the P branch to the left of the gap and $m = 1$ to 2906.2 cm^{-1} in the R branch to the right. Make other assignments consecutively working out from the center. Assignments for the entire spectrum are:

m	$\tilde{\nu}_P/\mathrm{cm}^{-1}$	m	$\tilde{\nu}_R/\mathrm{cm}^{-1}$
−1	2865.0	1	2906.2
−2	2843.5	2	2925.7
−3	2821.5	3	2944.8
−4	2798.9	4	2963.1
−5	2775.7	5	2981.0
−6	2751.8	6	2998.1
−7	2727.7	7	3014.5
−8	2702.9	8	3030.2
−9	2677.6	9	3045.1
−10	2651.8	10	3059.3

Example 5-2. Use the program `Linreg` to fit the $^1\mathrm{H}^{35}\mathrm{Cl}$ data obtained in the last example to Eq. (5.5), and calculate values for the rotational constants $\tilde{B}_0$ and $\tilde{B}_1$. Then calculate values of $\tilde{\alpha}_e$, $\tilde{B}_e$ and R_e for the molecule.

Answer. Enter data obtained from the table in Example 5-1 in `Linreg` beginning on the third line of code,

$$\text{Data} = \{\{-1, \ 2865.0\},$$
$$\{-2, \ 2843.5\},$$
$$\{-3, \ 2821.5\},$$
$$\text{etc.}$$

The program calculates

$$\tilde{\nu}_{00} = 2866 \ \mathrm{cm}^{-1}$$
$$\tilde{B}_1 + \tilde{B}_0 = 20.45 \ \mathrm{cm}^{-1}$$
$$\tilde{B}_1 - \tilde{B}_0 = -0.3027 \ \mathrm{cm}^{-1},$$

so

$$\tilde{B}_1 = \frac{20.45 - 0.3027}{2} = 10.07 \ \mathrm{cm}^{-1}$$
$$\tilde{B}_0 = 20.45 - 10.07 = 10.38 \ \mathrm{cm}^{-1}.$$

From Eq. (4.24) conclude that

$$\tilde{\alpha}_e = \tilde{B}_0 - \tilde{B}_1 = 0.303 \ \mathrm{cm}^{-1}$$
$$\tilde{B}_e = \frac{2\tilde{B}_0 + \tilde{\alpha}_e}{2} = \frac{(2)(10.38 \ \mathrm{cm}^{-1}) + (0.303 \ \mathrm{cm}^{-1})}{2} = 10.53 \ \mathrm{cm}^{-1}.$$

Calculate the reduced mass of ^{1}H^{35}Cl,

$$\mu = \frac{(1.0078 \text{ g mol}^{-1})(34.969 \text{ g mol}^{-1})}{(1.0078 \text{ g mol}^{-1} + 34.969 \text{ g mol}^{-1})(6.0221 \times 10^{23} \text{ mol}^{-1})}$$

$$= 1.6266 \times 10^{-24} \text{ g}$$

$$= 1.6266 \times 10^{-27} \text{ kg}.$$

Calculate I_e from Eq. (4.25),

$$I_e = \frac{h}{8\pi^2 \tilde{B}_e c}$$

$$= \frac{(6.6261 \times 10^{-34} \text{ J s})}{(8\pi^2)(10.53 \text{ cm}^{-1})(2.9979 \times 10^8 \text{ m s}^{-1})(100 \text{ cm m}^{-1})}$$

$$= 2.658 \times 10^{-47} \text{ kg m}^2.$$

Calculate R_e from Eq. (4.26),

$$R_e = \sqrt{I_e/\mu} = \left(\frac{2.658 \times 10^{-47} \text{ kg m}^2}{1.6266 \times 10^{-27} \text{ kg}}\right)^{1/2}$$

$$= 1.278 \times 10^{-10} \text{ m}$$

$$= 0.1278 \text{ nm}.$$

5.3 Vibrational–Electronic Spectroscopy

Here we neglect the rotational fine structure and calculate vibrational–electronic (vibronic) energy levels for diatomic molecules with Eqs. (4.30) and (4.31). Wavenumbers for transitions between an upper vibronic state defined by v', Λ' and a lower state defined by v'', Λ'' are calculated with

$$\tilde{v} = \frac{E(v', \Lambda') - E(v'', \Lambda'')}{hc}. \tag{5.6}$$

No selection rules restrict these transitions, and all values of $\Delta v = v' - v''$ are allowed. The program Vie12 calculates and plots vibrational–electronic spectra for diatomic molecules. It interprets a spectrum by sorting the peaks into sequences for $\Delta v = -3, -2, -1, 0, 1, 2, 3$. The program also calculates and prints a *Delandres table* whose entries $\tilde{T}(v', v'')$ are wavenumbers for the emission transitions $v', \Lambda' \rightarrow v'', \Lambda''$.

5.4   Rotational–Vibrational-Electronic Spectroscopy

Each peak plotted by the program `Viel2` (Sec. 5.3) is, in fact, a band of peaks with a rotational fine structure. The program `Roviel` expands `Viel2` to calculate and plot this fine structure. Energy equations involved (for diatomic molecules) are (4.33) and (4.34), and the wavenumber for a transition $J', v', \Lambda' \leftarrow J'', v'', \Lambda''$ is calculated with

$$\tilde{\nu} = \frac{E(J', v', \Lambda') - E(J'', v'', \Lambda'')}{hc}.$$

The program isolates the rotational contributions by first calculating a vibrational term,

$$\tilde{T}_0 = \tilde{T}'_e + \tilde{\omega}'_e(v' + 1/2) - \tilde{\omega}'_e x'_e(v' + 1/2)^2$$
$$- \tilde{T}''_e - \tilde{\omega}''_e(v'' + 1/2) + \tilde{\omega}''_e x''_e(v'' + 1/2)^2 \tag{5.7}$$

and adding to this rotational terms permitted by the following selection rules:

If $\Delta\Lambda = 0$ and $\Lambda'' = \Lambda' = 0$, $\Delta J = \pm 1$ and the band has P and R branches but no Q branch.

If $\Delta\Lambda = \pm 1$, $\Delta J = 0, \pm 1$ and the band has a strong Q branch, as well as P and R branches.

If $\Delta\Lambda = 0$ and $\Lambda'' = \Lambda' \neq 0$, $\Delta J = 0, \pm 1$ and the band has a weak Q branch, as well as P and R branches.

`Roviel` plots lines for the 4-11 band (i.e., $v' = 4, v'' = 11$) of the CO(g) $A^1 \prod - X^1 \sum^+$ spectrum. Run the program and sort out the details by plotting the P, Q, and R branches separately.

5.5   Magnetic Resonance Spectroscopy

Spin states of electrons and nuclei have slightly different energies in an applied magnetic field. In the magnetic resonance method of spectroscopy absorption transitions between electronic or nuclear spin states are induced by radio-frequency photons.

NMR

Nuclear magnetic resonance (NMR) spectroscopy represents magnetically different nuclei with *multiplets*. Each multiplet can be interpreted to measure the number of equivalent nuclei involved, the number of coupling interactions those nuclei have with other nuclei, and the energies of the interactions.

The program Nmr provides a first-order simulation of the NMR multiplet pattern for a given system of spin-1/2 nuclei. The next example shows how to use the program and interpret its message.

Example 5-3. Use the program Nmr to simulate the NMR multiplet pattern for a ^{1}H spin system of the AMX kind. Coupling constants are $J_{AM} = 50$ Hz, $J_{AX} = 20$ Hz, and $J_{MX} = 100$ Hz. Chemical-shift frequencies for the three nuclei are $\nu_A = 200$ Hz, $\nu_M = 500$ Hz, and $\nu_X = 800$ Hz.

Answer. Enter the A, M, and X frequencies, in the list f0

$$f0 \ = \ \{200., \ 500., \ 800.\};$$

Enter the number of equivalent nuclei for each multiplet in the list ne,

$$ne \ = \ \{1, \ 1, \ 1\};$$

Enter the number of coupling interactions for each nucleus in the list ni,

$$ni \ = \ \{2, \ 2, \ 2\};$$

Enter the coupling constants for each multiplet in the list J in descending order,

$$J \ = \ \{\{50., \ 20.,\}, \ \{100., \ 50.\}, \ \{100., \ 20.\}\};$$

Finally, in the list y0 enter values of the parameters for locating vertically the branching diagrams that interpret the multiplets,

$$y0 \ = \ \{3., \ 3., \ 3.\};$$

Run the program and note the double-doublet pattern for each multiplet. The branching diagrams locate the centers of the multiplets and display splittings for each coupling interaction.

ESR

Electron magnetic resonance or *electron spin resonance (ESR) spectroscopy* provides the same kind of information as NMR for systems involving unpaired electron spins. The program Esr simulates multiplet patterns for a single electron coupling with spin-1/2 nuclei. The program is similar to Nmr in the way it runs and is interpreted.

2D NMR

Complicated NMR spectra can become almost indeciperable when the multiplets are extensively overlapped or superimposed. A method called *two-*

dimensional NMR expands the spectrum into two dimensions and greatly simplifies the task of deducing coupling interactions. Each nucleus has its own spectrum displayed separately in a row (or column) of the second dimension. The program 2dnmr illustrates by simulating two-dimensional ^{1}H spectra. The next example introduces the program.

Example 5-4. In a one-dimensional NMR spectrum peaks representing five magnetically different ^{1}H nuclei are located at the delta values 2.52, 3.86, 4.58, 5.02, and 6.15 ppm. Coupling interactions among these ^{1}H nuclei are represented by the following triangular table:

	1	2	3	4	5
1	5.1	2.5	0	0	2.2
2		4.2	1.9	2.8	0
3			6.5	2.0	0
4				5.4	0
5					5.2

The first number in the first row is the intensity of ^{1}H nucleus 1 (with the lowest chemical shift) in the 1D spectrum, the second the intensity of the coupling between ^{1}H nucleus 1 and ^{1}H nucleus 2 (with the second lowest chemical shift), the third the intensity of the coupling between ^{1}H nucleus 1 and ^{1}H nucleus 3, and so forth. The second row contains data for ^{1}H nucleus 2 in the 1D spectrum and for coupling between ^{1}H nucleus 2 and ^{1}H nuclei 3, 4, ... The remaining rows contain data of the same kind for ^{1}H nuclei 3, 4, ... Enter these data in the program 2dnmr, run the program and correlate the display with the data.

Answer. Enter delta values in ascending order in the list delta,

    delta = {2.52, 3.86, 4.58, 5.02, 6.15};

And then coupling data from the above table in the list data,

    data = {{5.1, 2.5, 0, 0, 2.2},

        {4.2, 1.9, 2.8, 0},

        {6.5, 2.0, 0},

        {5.4, 0},

        {5.2}};

Run the program and note the 5 peaks as delta values plotted on the diagonal of the two-dimensional display. Also note the *cross peaks*, each one indicating a coupling interaction. For example, the red circle located by the coordinates 2.52, 3.86 represents the coupling interaction between ^{1}H nuclei 1 and 2.

5.6   Fourier-Transform Methods

The classical approach to spectrometer design is to spread a radiation beam into spectral elements each covering a narrow wavenumber range. Good spectrometers can achieve high resolution, but the dispersion has disadvantages. Each spectral element carries a very small amount of energy, thus limiting the sensitivity of the instrument, and the rate of scanning of a full spectrum is inconveniently slow. The Fourier-transform method introduces a radical departure: it permits the detector to scan an entire spectrum without dispersion. It is capable of high resolution and can gather and process the information needed to generate a spectrum in seconds compared to minutes for dispersion instruments.

The program Irft demonstrates the Fourier-transform technique applied to infrared spectroscopy. The program simulates the *interferogram* recorded by an infrared spectrometer and then calculates and plots the Fourier transform of the interferogram. The transform displays the spectral peaks.

At a mirror displacement of d (in cm) the interferogram is simulated with the function

$$I(d) = \sum_i a_i \cos(2\pi\tilde{v}_i d)\exp(-2\pi b_i|d|), \qquad (5.8)$$

in which $\tilde{v}_i$ is the wavenumber (in cm^{-1}) for the ith spectral peak, b_i is the width of the peak (in cm^{-1}), and a_i is an amplitude factor. Values of a_i, b_i, and $\tilde{v}_i$ for the spectral peaks are entered in Irft in the lists pa, pb, and pw. The program samples the interferogram function (5.8) k times between the mirror displacements -dMax and dMax, uses these data to calculate the Fourier transform and thus display the spectral peaks. The ith peak is centered at the wavenumber $\tilde{v}_i$, has the width b_i, and the amplitude $a_i/(2\pi b_i)$. As the program demonstrates, the Fourier-transform analysis cannot, without some confusion, locate peaks whose wave-numbers are larger than k/(2 dMax). An *aliasing* error intrudes, which is discussed further in Exercise 5-21.

The program Nmrft simulates the application of Fourier-transform methods to NMR spectroscopy. In the NMR case an *FID signal* is recorded by the instrument and its Fourier transform displays the spectral peaks. Nmrft is formally similar to Irft. At a time t (in s) the FID signal is simulated by the function

$$I(t) = \sum_i a_i \cos(2\pi v_i t)\exp(-2\pi b_i t), \qquad (5.9)$$

in which v_i is the frequency of the ith spectral peak (in s^{-1}), b_i is the width of the peak (in s^{-1}), and a_i is an amplitude factor. Values of a_i, b_i, and v_i for the spectral peaks are entered in Nmrft in the lists pa, pb, and pf. The program samples the FID function (5.9) k times between the times 0 and tMax and uses these data to calculate the Fourier transform and thus display the spectral peaks. The ith peak is centered at the frequency v_i, has the width b_i and amplitude $a_i/(4\pi b_i)$. Here confusion results if the Fourier-transform analysis attempts to locate peaks whose frequencies are larger than $k/(2\,tMax)$.

5.7   Exercises

Rotational Spectroscopy

5-1   Use the program Ro3 to calculate and plot the rotational spectrum for $^1H^{35}Cl(g)$ in its ground vibrational and electronic state at 300 K. Rotational and vibrational constants for $^1H^{35}Cl$ are

$$\tilde{B}_e = 10.59342 \text{ cm}^{-1}$$

$$\tilde{\alpha}_e = 0.30718 \text{ cm}^{-1}.$$

These data and others quoted for diatomic molecules come from Huber and Herzberg, and Herzberg, *Spectra of Diatomic Molecules* (1950).

5-2   Use the program Ro3 to calculate and plot the rotational spectrum for $^9Be^{16}O(g)$ in its ground vibrational and electronic state at 300 K. Rotational and vibrational constants for $^9Be^{16}O$ are

$$\tilde{B}_e = 1.6510 \text{ cm}^{-1}$$

$$\tilde{\alpha}_e = 0.0190 \text{ cm}^{-1}.$$

5-3   Use the program Ro3 to calculate and plot the rotational spectrum for $^{63}Cu^2H(g)$ in its ground vibrational and electronic state at 300 K. Rotational and vibrational constants for $^{63}Cu^2H$ are

$$\tilde{B}_e = 4.0375 \text{ cm}^{-1}$$

$$\tilde{\alpha}_e = 0.09140 \text{ cm}^{-1}.$$

Rotational–Vibrational Spectroscopy

5-4   Use the program Rovi2 to calculate and plot the rotational–vibrational spectrum for $^1H^{35}Cl(g)$ in its ground electronic state at 300 K. Use data given in Exercise 5-1 and

$$\tilde{\omega}_e = 2990.946 \text{ cm}^{-1}$$

$$\tilde{\omega}_e x_e = 52.819 \text{ cm}^{-1}.$$

5-5   Use the program Rovi2 to calculate and plot the rotational–vibrational spectrum for $^9Be^{16}O(g)$ in its ground electronic state at 300 K. Use data given in Exercise 5-2 and

$$\tilde{\omega}_e = 1487.323 \text{ cm}^{-1}$$

$$\tilde{\omega}_e x_e = 11.8297 \text{ cm}^{-1}.$$

5-6   Use the program Rovi2 to calculate and plot the rotational–vibrational spectrum for $^{63}Cu^2H(g)$ in its ground electronic state at 300 K. Use data given in Exercise 5-3 and

$$\tilde{\omega}_e = 1384.38 \text{ cm}^{-1}$$

$$\omega_e x_e = 19.14 \text{ cm}^{-1}.$$

5-7   Use the program Symtop1 to plot a parallel band in the rotational–vibrational spectrum for $^{14}N^1H_3(g)$ whose lower and upper rotational constants are (ignoring the effects of inversion):

$$\tilde{B}_0 = 9.4443 \text{ cm}^{-1}$$

$$\tilde{C}_0 = 6.196 \text{ cm}^{-1}$$

$$\tilde{B}_1 = 9.356 \text{ cm}^{-1}$$

$$\tilde{C}_1 = 6.178 \text{ cm}^{-1}.$$

The center of the band is located at 3336 cm^{-1}. These data and those quoted in the next exercise are obtained from Herzberg, *Infrared and Raman Spectra* (1945).

5-8   Use the program Symtop2 to plot a perpendicular band in the rotational-vibrational spectrum for $^{14}N^1H_3(g)$ whose lower and upper rotational constants are (ignoring the effects of inversion):

$$\tilde{B}_0 = 9.4443 \text{ cm}^{-1}$$

$$\tilde{C}_0 = 6.196 \text{ cm}^{-1}$$

$$\tilde{B}_1 = 9.674 \text{ cm}^{-1}$$

$$\tilde{C}_1 = 6.130 \text{ cm}^{-1}.$$

The center of the band is located at 1626 cm^{-1}.

Data Analysis

5-9   In Example 5-1 we used the program Peaks to define the $^1H^{35}Cl$ part of the HCl rotational–vibrational spectrum. Then in Example 5-2 we used the program Linreg to fit the data to the empirical Eq. (5.5), to calculate $\tilde{B}_0$ and $\tilde{B}_1$, and finally to calculate R_e for the molecule. Do this same analysis for the $^1H^{37}Cl$ data obtained by Peaks. How does the R_e value for $^1H^{37}Cl$ compare with that for $^1H^{35}Cl$?

5-10 In Example 5-2 we calculated one important parameter for HCl, its interatomic distance R_e. We go further now and calculate the force constant k for the molecule. As preparation for that calculation derive the equation

$$\tilde{v}_{00}(v) = v\tilde{\omega}_e - \tilde{\omega}_e x_e v(v+1), \qquad (5.10)$$

which locates centers of the bands for the $v \leftarrow 0$ transitions.

5-11 Data are quoted below for centers of the fundamental and overtone bands (i.e., for the transitions $v \leftarrow 0$ with $v = 1, 2, \ldots$) observed for $^1H^{35}Cl(g)$. Use the program Linreg to fit these data to Eq. (5.10) and calculate a value for $\tilde{\omega}_e$. Then calculate the force constant k.

v	$\tilde{v}_{00}(v \leftarrow 0)/\text{cm}^{-1}$
1	2885.9
2	5668.0
3	8347.0
4	10923.1
5	13396.5

Vibrational–Electronic Spectroscopy

5-12 Use the program Viel2 and the data quoted in Exercise 4-18 to plot the corresponding vibrational–electronic spectrum for the $^{14}N_2(g)$ electronic states $A^3 \Sigma_u^+$ and $B^3 \Pi_g$.

5-13 Use the program Viel2 and the data quoted below for the $^{14}N_2(g)$ electronic states $C^3 \Pi_u$ and $B^3 \Pi_g$ to plot the corresponding vibrational–electronic spectrum.

$$\tilde{\omega}_e'' = 1733.39 \text{ cm}^{-1}$$

$$\tilde{\omega}_e'' x_e'' = 14.122 \text{ cm}^{-1}$$

$$\tilde{T}_e'' = 59619.4 \text{ cm}^{-1}$$

$$\tilde{\omega}_e' = 2047.18 \text{ cm}^{-1}$$

$$\tilde{\omega}_e' x_e' = 28.445 \text{ cm}^{-1}$$

$$\tilde{T}_e' = 89136.88 \text{ cm}^{-1}.$$

Rotational–Vibrational–Electronic Spectroscopy

5-14 Use the program `Roviel` with data quoted below for the $^9Be^{16}O(g)$ electronic states $B^1\sum^+$ and $X^1\sum^+$ to plot rotational lines for the 0-0 band at 300 K.

$$\tilde{\omega}''_e = 1487.323 \text{ cm}^{-1}$$

$$\tilde{\omega}''_e x''_e = 11.8297 \text{ cm}^{-1}$$

$$\tilde{\alpha}''_e = 0.0190 \text{ cm}^{-1}$$

$$\tilde{B}''_e = 1.6510 \text{ cm}^{-1}$$

$$\tilde{T}''_e = 0$$

$$\tilde{\omega}'_e = 1370.817 \text{ cm}^{-1}$$

$$\tilde{\omega}'_e x'_e = 7.7455 \text{ cm}^{-1}$$

$$\tilde{\alpha}'_e = 0.0154 \text{ cm}^{-1}$$

$$\tilde{B}'_e = 1.5758 \text{ cm}^{-1}$$

$$\tilde{T}'_e = 21253.94 \text{ cm}^{-1}.$$

5-15 Use the program `Roviel` with data quoted below for the $^{63}Cu^2H(g)$ electronic states $A^1\Sigma^+$ and $X^1\Sigma^+$ to plot rotational lines for the 0-0 band at 1000 K.

$$\tilde{\omega}''_e = 1384.38 \text{ cm}^{-1}$$

$$\tilde{\omega}''_e x''_e = 19.14 \text{ cm}^{-1}$$

$$\tilde{\alpha}''_e = 0.0914 \text{ cm}^{-1}$$

$$\tilde{B}''_e = 4.0375 \text{ cm}^{-1}$$

$$\tilde{T}''_e = 0$$

$$\tilde{\omega}'_e = 1213.16 \text{ cm}^{-1}$$

$$\tilde{\omega}'_e x'_e = 20.65 \text{ cm}^{-1}$$

$$\tilde{\alpha}'_e = 0.0898 \text{ cm}^{-1}$$

$$\tilde{B}'_e = 3.5199 \text{ cm}^{-1}$$

$$\tilde{T}'_e = 23412 \text{ cm}^{-1}.$$

Magnetic Resonance Spectroscopy

5-16 Simulate NMR multiplets for a ^1H spin system of the A_3X kind using the program `Nmr`. The coupling constant is $J_{AX} = 50$ Hz and chemical-shift frequencies for the two nuclei are $v_A = 200$ Hz and $v_X = 800$ Hz.

5-17 Simulate NMR multiplets for a ^{1}H spin system of the A_2X_2 kind using the Nmr program. The coupling constant is $J_{AX} = 5$ Hz. Chemical-shift frequencies for the two nuclei are $\nu_A = 200$ Hz and $\nu_X = 800$ Hz.

5-18 Simulate the ESR spectrum for the $CH_2OH_5\bullet$ radical using the program Esr. Coupling constants between the unpaired electron and the methylene ^{1}H and hydroxyl ^{1}H are 17.46 and 1.15 Gauss. Use the derivative plotting option and assume an arbitrary value of zero for the frequency at the center of the multiplet.

5-19 Simulate the ESR spectrum for the ethyl radical $C_2H_5\bullet$ using the program Esr. Coupling constants between the unpaired electron and the methyl ^{1}H and methylene ^{1}H are 26.87 and 22.38 Gauss. The center of the multiplet is located at 3294 Gauss. Interpret the branching diagram. Notice that some of the branches overlap.

5-20 A 2D NMR spectrum for ^{1}H nuclei in N-methyl-benzocarbostyril is simulated in the file Ex5-20. Run the program and interpret the spectrum. Approximate chemical shifts for the nuclei and their positions in the molecule are

Position	δ/ppm
5	8.08
6	7.62
7	7.48
8	8.45
11	8.08
12	7.21
13	7.42
14	7.25

Fourier-Transform Methods

5-21 Fourier-transform calculations are limited by an error called *aliasing*, which causes confusion in applications of the program Irft if wavenumbers included in the interferogram exceed k/(2 dMax), where k is the number of points sampled in the Fourier-transform calculation and dMax is the maximum mirror displacement. For example, if k = 500 and dMax = 0.05 then wavenumbers contributing to the interferogram should not exceed $500/0.1 = 5000$ cm^{-1}. Note what happens if this rule is violated by running Irft with peaks at 1000, 2000, and 6000 cm^{-1} as input. How can this problem be remedied?

5-22 The last exercise illustrated the aliasing error the Fourier-transform calculation can make if an interferogram is not sampled with enough points.

Another kind of sampling error results when the interferogram is truncated on the sides. When this error is serious, the Fourier-transform calculation produces peaks that are distorted in width and height. Demonstrate by running Irft for a single peak whose width is $pb = 5$ cm^{-1} and is centered at 1000 cm^{-1}. Use $pa = 20$ for the interferogram amplitude and set the number of sampling points at $k = 500$. In this case the peak height produced by the Fourier-transform calculation should be $20/(5)(2\pi) = 0.637$. Instead, the program calculates 0.504. What evidently is the remedy for *this* problem?

6

Solids, Liquids, and Surfaces

So far in this part of the book, we have pretended that molecules are small in size (containing no more than a few atoms) and in the aggregate are found only in gaseous phases. It is time now to recognize several important further facts of molecular life—that molecules are also found in solid, liquid, and surface phases, and that they can be large and very large, perhaps containing thousands of atoms. In this chapter we see molecules in solids, in liquids, and on surfaces. The next chapter discusses large-sized molecules (macromolecules).

X-ray diffraction methods for studying the crystalline solid state are demonstrated in Sec. 6.1 (programs `Pattersn`, `Powder`, `Xray1`, `Xray2`, `Xray3` and `Xray4`). A similar diffraction method utilizing electron beams is the topic in Sec. 6.2 (the program `Elecdiff`). In Sec. 6.3 electrical properties of solids are the subject and a program (`Fermi`) is introduced that calculates concentrations of current carriers in semiconductors. A collection of QuickBASIC programs (MC1.BAS, MC2.BAS, MD1.BAS, MD2.BAS, and MD3.BAS), introduced in Secs. 6.4 and 6.5, shows how molecular dynamics and Monte Carlo calculational methods are applied to the determination of liquid structures. Finally, surface structures are emphasized, electrical aspects in Sec. 6.6 (the programs `Gouy1`, `Gouy2`, and `Gouy3`) and surface crystallography in Sec. 6.7 (the program `Leed`).

6.1 X-Ray Crystallography

X rays are reflected by lattice planes in a crystalline solid when they bounce off the planes with the same angle of reflection as incidence. The effect is observable if all the reflections from a set of planes are in phase so they can reinforce each other. That condition is met if all the reflected rays travel distances that differ by an integer number of wavelengths λ. The Bragg equation,

$$2d \sin \theta = \lambda, \qquad (6.1)$$

with d the perpendicular distance between the reflecting planes, and θ the angle of reflection, assures the conditions of reinforcement.

A crystal lattice is characterized by a unit cell, which generates the entire lattice when it is translated in three dimensions. Reflecting planes in the lattice are designated by the Miller indices h, k, and l, and if the unit cell has rectangular axes, and its dimensions are a, b, and c, the interplanar distances d are determined by

$$\left(\frac{1}{d}\right)^2 = \left(\frac{h}{a}\right)^2 + \left(\frac{k}{b}\right)^2 + \left(\frac{l}{c}\right)^2,$$

or

$$Q = Ah^2 + Bk^2 + Cl^2, \tag{6.2}$$

in which $Q = 1/d^2$, $A = 1/a^2$, $B = 1/b^2$ and $C = 1/c^2$. The x-ray wavelength λ and the Bragg angle θ are both measurable, so for each reflection a value of $Q = 1/d^2$ can be calculated using Bragg's equation (6.1).

One of the standard problems in x-ray crystallography is to observe as many reflections as possible from a crystalline material, then to determine Miller indices for each reflection, and finally to use this information to calculate the lattice parameters a, b, and c, and others if the unit cell does not have rectangular axes.

The Powder Method

A commonly used technique for performing this task, called the powder method, was discovered by Debye, Scherrer, and Hull. As the name implies, the sample for this kind of experiment is provided in finely powdered form, which means that many different crystallites are presented to the x-ray beam in many different orientations. As the angle of incidence of the x-ray beam is changed many reflections are recorded.

The program `Powder` does part of the analysis of data obtained this way. It begins by reading a file of data for interplanar distances d calculated according to Bragg's equation (6.1) from observed reflection angles θ. Then, with further information supplied on the unit-cell dimensions a, b, and c, the program indexes each reflection by assigning to it Miller indices h, k, and l for the reflecting planes.

The Precession Method

In this technique, a single-crystal sample is mounted with one of the axes of its lattice, let us say the z axis, perpendicular to a photographic plate or other detector. Reflections are recorded as spots by the detector, and they are arranged in a grid. Miller indices for the set of reflecting planes responsible for a spot are easily read directly from the spot's location in the grid.

The spots in an x-ray diffraction pattern have different intensities that depend on locations of the atoms in the molecules forming the lattice. Analysis of these intensities provides detailed information on the molecular structure. One key to that analysis is a quantity called a structure factor, often defined

$$F_{hkl} = \sum_i f_i \cos[2\pi(hX_i + kY_i + lZ_i)]. \tag{6.3}$$

The summation covers all of the atoms in the unit cell. Coordinates of the atoms in the unit cell are expressed as reduced coordinates X_i, Y_i, and Z_i, each of which is an actual coordinate x_i, y_i, or z_i divided by the corresponding lattice parameter

$$X_i = x_i/a, \quad Y_i = y_i/b, \quad \text{and} \quad Z_i = z_i/c. \tag{6.4}$$

The factors f_i in Eq. (6.3), called scattering factors, depend on the type of atom involved and also on the Bragg angle θ, the x-ray wavelength λ, and the temperature.

Structure factors are important for two reasons. First, the magnitudes of structure factors are proportional to the measurable spot intensities, that is, $|F_{hkl}|$ is proportional to the intensity of the spot identified with the Miller indices h, k, and l. Second, the set of structure factors F_{hkl} for an entire diffraction pattern can be used to calculate the electron density $\rho(X, Y, Z)$ at any location in the unit cell, with X, Y, and Z again representing reduced coordinates. If Eq. (6.3) is valid, the equation that calculates $\rho(X, Y, Z)$ is a Fourier expansion with the structure factors F_{hkl} as Fourier coefficients,

$$\rho(X, Y, Z) = (2/V) \sum_{h,k,l} F_{hkl} \cos[2\pi(hX + kY + lZ)], \tag{6.5}$$

where V is the volume of the unit cell, and the summation covers all of the spots in the diffraction pattern. Equations (6.3) and (6.5) are valid only for structures with centrosymmetric unit cells, those having symmetry around the center: for each atom located at X, Y, Z in the unit cell there is also one located at $1 - X, 1 - Y, 1 - Z$.

Structure factors for lattices with noncentrosymmetric unit cells are like vectors with two components. One vector component, written A_{hkl}, is a sum of cosine terms like the sum in Eq. (6.3),

$$A_{hkl} = \sum_i f_i \cos[2\pi(hX_i + kY_i + lZ_i)], \tag{6.6}$$

and the other component, B_{hkl}, is a sum of sine terms,

$$B_{hkl} = \sum_i f_i \sin[2\pi(hX_i + kY_i + lZ_i)]. \tag{6.7}$$

The two components A_{hkl} and B_{hkl} determine the magnitude of the vector-like structure factor,

$$|F_{hkl}| = (A_{hkl}^2 + B_{hkl}^2)^{1/2}. \tag{6.8}$$

The noncentrosymmetric problem is more complicated than the centrosymmetric. To define a structure factor F_{hkl} you must determine its components, both A_{hkl} and B_{hkl}, or the magnitude $|F_{hkl}|$ and the ratio B_{hkl}/A_{hkl}, conventionally expressed as a *phase angle* ϕ_{hkl},

$$\phi_{hkl} = \tan^{-1}(B_{hkl}/A_{hkl}). \tag{6.9}$$

Both of the components are needed in the calculation of the electron density as a Fourier expansion,

$$\rho(X, Y, Z) = (1/V) \sum_{h,k,l} \{A_{hkl} \cos[2\pi(hX + kY + lZ]$$
$$+ B_{hkl} \sin[2\pi(hX + kY + lZ)]\}. \tag{6.10}$$

All of this is demonstrated by the programs Xray1 and Xray2 for the centrosymmetric and noncentrosymmetric cases defined in two dimensions. Atomic coordinates X_i, Y_i, and the x-ray wavelength λ are supplied to the programs. They simulate the diffraction spot pattern against an h, k grid, and then calculate electron densities $\rho(X, Y)$, which are displayed as a contour map and also as a three-dimensional surface. Run both programs with the atomic coordinates provided.

Patterson Plots

Once the structure factors F_{hkl} are fully determined they solve the structure, that is, in Eq. (6.5) or (6.10) they calculate an electronic density map of the unit cell and the lattice. But there is a catch. The x-ray diffraction experiment measures only magnitudes of the structure factors from intensities of the reflections. Structure factors also have signs in structures with centrosymmetric unit cells and phase angles in structures with noncentrosymmetric unit cells. These aspects of structure factors are not directly measurable. In other words, the crucial structure factors are only about half determined by x-ray diffraction data. This is the ubiquitous phase problem that creates most of the calculational difficulty of x-ray crystallography.

The unwelcome obstacle of the phase problem has been surmounted in many ingenious ways. An analysis developed by Patterson is one of them. Patterson's strategy is to calculate a Patterson function $P(X, Y, Z)$ with equations like (6.5) and (6.10) except that F_{hkl} is replaced by the squared magnitude $|F_{hkl}|^2$. The Patterson counterpart of Eq. (6.5) is

$$P(X, Y, Z) = (1/V) \sum_{h,k,l} |F_{hkl}|^2 \cos[2\pi(hX + kY + lZ)]. \tag{6.11}$$

No phases or signs are needed to determine the $|F_{hkl}|^2$ factors in this summation, so the Patterson function can be calculated and plotted directly from the diffraction data.

When the Patterson function $P(X, Y, Z)$ is mapped on the same scale as the unit cell, it generates peaks which measure (from the corners of the map) all of the interatomic distances and their directions in the unit cell. If there are N atoms in the unit cell there are $N(N - 1)$ peaks on this Patterson map, not including the large peaks that always appear at the corners of the map.

Two programs, Xray3 and Xray4, calculate and plot Patterson maps from structure factors in much the same way Xray1 and Xray2 plot and calculate electron density maps. The program Pattersn, introduced in Exercise 6-8, sketches an idealized Patterson map and shows how it relates to locations of atoms in the unit cell.

6.2   Electron Diffraction in Gases (an Orphan Topic)

In the last section we were concerned with x-ray diffraction produced by lattices of molecules in crystals. If the molecules are present in a gas phase at low pressures no lattice forms and no diffraction effects involving lattice planes, or any other kind of *inter*molecular interference, is observed. *Intra*molecular interference effects can, however, be observed. The diffraction patterns obtained resemble the patterns observed when solid powder samples are used because the gas-phase molecules, like the crystallites in the powder, present all possible orientations to the x-ray beam. We are straying here from the themes of this chapter, solids, liquids, and surfaces, but the link with diffraction methods justifies the misplacement.

Using an argument that bears some resemblance to that leading to Bragg's equation (6.1), an approximate equation for the intensity I of an observable (reinforced) reflection of an electron beam by a gas-phase molecule is derived,

$$I = k \sum_{i \neq j} \frac{Z_i Z_j \sin(sR_{ij})}{sR_{ij}}, \qquad (6.12)$$

in which Z_i and Z_j are nuclear charges for the ith and jth atoms in the molecule, R_{ij} is the distance between the same two atoms. The scattering variable s in Eq. (6.12) is an observable quantity determined by the wavelength λ of the electron beam and the angle of reflection θ,

$$s = \frac{4\pi \sin(\theta/2)}{\lambda}. \qquad (6.13)$$

The program Elecdiff applies Eq. (6.12) to $SiCl_4$ and fits the equation to electron-diffraction data. The program finally calculates a value of R_{SiCl} for the Si-Cl bond distance. Run the program and note the strategy for matching minimum and maximum observed values of s with the minima and maxima generated by Eq. (6.12).

6.3   Semiconductors

Electrons in semiconductors occupy states available in the *valence band* of the solid state, and to a lesser extent states in the *conduction band*. In a *pure-material semiconductor* the *energy gap* between the valence band and the conduction band is forbidden territory for electron occupation. Semiconductors can be modified, however, with controlled amounts of impurities in the lattice which make states available for electron occupation in the otherwise forbidden energy gap.

An electron promoted to the conduction band becomes an electrical current carrier in the conduction band and it leaves behind a *hole*, which serves as a carrier in the valence band. Electrons and holes can also be trapped at *donor* and *acceptor impurity states* located in the energy gap.

Semiconductor equilibrium calculations recognize six electron and hole concentrations:

$[e^-]_c$ = electron concentration in the conduction band

$[e^-]_d$ = electron concentration in a donor impurity state

$[e^-]_a$ = electron concentration in an acceptor impurity state

$[h^+]_v$ = hole concentration in the valence band

$[h^+]_d$ = hole concentration in a donor impurity state

$[h^+]_a$ = hole concentration in an acceptor impurity state.

Four of these concentrations are related by

$$[h^+]_v + [h^+]_d = [e^-]_c + [e^-]_a. \tag{6.14}$$

The condition $[e^-]_c > [h^+]_v$ defines the case of an *n-type semiconductor* and $[e^-]_c < [h^+]_v$ a *p-type semiconductor*.

The crucial concentrations $[e^-]_c$ and $[h^+]_v$ are calculated with

$$[e^-]_c = \frac{2N_c}{\exp[(E_g - E_f)/k_B T] + 1} \tag{6.15}$$

and

$$[h^+]_v = \frac{2N_v \exp[(E_v - E_f)/k_B T]}{\exp[(E_v - E_f)/k_B T] + 1}, \tag{6.16}$$

in which N_c and N_v equal total concentrations of states in the conduction and valence bands, E_g is the energy level at the top of the energy gap (or the bottom of the conduction band), E_v is the energy level at the top of the valence band, and E_f is a useful parameter called the *Fermi level*.

Equations for the concentrations $[e^-]_a$ and $[h^+]_d$ appearing in Eq. (6.14) are

$$[e^-]_a = \frac{[A]_0}{\exp[(E_a - E_f)/k_B T] + 1} \tag{6.17}$$

$$[h^+]_d = \frac{[D]_0 \exp[(E_d - E_f)/k_B T]}{\exp[(E_d - E_f)/k_B T] + 1}, \tag{6.18}$$

where $[A]_0$ and $[D]_0$ are concentrations of acceptor and donor states, and E_a and E_d are energy levels for those states.

With data supplied for $E_v, E_g, E_a, E_d, [A]_0$ and $[D]_0$, the program Fermi solves Eqs. (6.14) to (6.18) simultaneously to obtain a value for the Fermi level E_f, and then uses this result to calculate all of the relevant electron and hole concentrations.

6.4 Molecular Dynamics Simulations of Liquids

The methods of molecular dynamics simulate macroscopic systems (liquid and otherwise) by focusing on only a few molecules in a small part of the system called a *simulation box*. In our examples the box is two dimensional and square, or three dimensional and cubic, and it contains no more than 50 molecules. Two conventions permit the enormous extrapolation from the tiny box to the full macroscopic system:

- *The periodic-boundary convention.* We assume that the simulation box is part of a periodic system of replicas of the box. Thus a molecule leaving the box by crossing a boundary is replaced by a molecule crossing the opposite boundary.
- *The minimum-image convention.* Molecular models of liquids usually include interactions among molecules. When these calculations are done for a particular molecule they obviously cannot include all of the other molecules in the macroscopic system (about 10^{24}!). The interactions are conveniently truncated by placing the molecule of interest at the center of a box of the same size as the simulation box, and assuming that the molecule interacts only with other molecules in that box.

In a molecular-dynamics simulation, paths of molecules in the simulation are followed by solving the classical equations of motion. The simulation begins by putting molecules in a simulation box of volume V and giving them initial velocities which are random but adjusted so the total momentum for all the molecules in the box is equal to zero. The molecules move in straight lines until a collision between two molecules occurs. Then the velocity components are changed for the colliding molecules and positions of all other molecules are updated. The molecules are again allowed to move until a collision occurs, requiring changes in velocity components and positions, and so forth.

The QuickBASIC program MD1.BAS implements this calculation. The program initially places the molecules in a face-centered cubic lattice and follows their subsequent trajectories as the dynamics evolve. The calculation is simplified by assuming that each side L of the cubic simulation box is equal to one, so the average molecular density ρ in a cubic box of volume V containing N molecules is $\rho = N/V = N/L^3 = N$. The looseness of molecular packing in the box is determined by ρ and also by the molecular diameter σ. We define a dimensionless parameter $\rho^* = \rho\sigma^3$, called the *reduced density*, which includes both of these quantities. Packing in the box is also expressed by the ratio V/V_0, where V_0 is the volume of a box that holds the molecules in *close packing*, so they cannot move at all. The reduced density ρ^* is related to V/V_0 according to

$$V/V_0 = \sqrt{2}/\rho^*. \qquad (6.19)$$

With V/V_0 set at values larger than about 1.58 MD1.BAS simulates the behavior of a liquid. Molecular motion is displayed in MD1.BAS by plotting positions of the centers of all the molecules in the box projected into two dimensions on one face of the box.

The QuickBASIC program MD2.BAS does a two-dimensional version of the molecular dynamics calculation, showing the erratic motion of a particular molecule as it bumps, and is bumped by, its neighbors. A third program, MD3.BAS, uses the two-dimensional calculation to plot a *radial distribution function* $g(r)$, which shows how neighboring molecules are distributed around a typical molecule. See the comment preceding Exercise 6-17 for information on how to run these QuickBASIC programs.

6.5   Monte Carlo Simulations of Liquids

This method also makes use of the simulation box and its conventions, but it simulates molecular motion in a different way. The strategy of the Monte Carlo method for "hard" molecules is to move one molecule in the system randomly and then check the resulting configuration for overlaps. If there are any, the move is rejected and another tried. This process is repeated for all the molecules in the simulation box. Then another such calculational cycle is carried out, and so forth. The QuickBASIC program MC1.BAS implements this procedure in two dimensions and calculates the radial distribution function $g(r)$.

The strategy of the Monte Carlo method for "soft" molecules is similar except that it allows molecules to penetrate each other to the extent allowed by an intermolecular potential energy function. We use the *Lennard–Jones potential energy function*,

$$V(r) = 4\varepsilon \left[\left(\frac{r}{\sigma}\right)^{-12} - \left(\frac{r}{\sigma}\right)^{-6} \right], \qquad (6.20)$$

in which ε and σ are parameters that are characteristic of the molecules involved. Note that ε has energy units and σ length units. In this case, the Monte Carlo procedure is to move a molecule randomly and then calculate the potential energy between that molecule and all the others. A move that decreases the potential energy is accepted; a move that increases the potential energy is accepted with the Boltzmann probability factor $\exp(-\Delta V/k_B T)$, in which ΔV is the potential energy increase.

The Monte-Carlo calculation with a potential energy function includes the temperature T as a relevant variable. For convenience, the calculation is done with temperature units ε/k_B. This dimensionless *reduced temperature* T^* is related to T in Kelvin units according to

$$T^* = \frac{T}{\varepsilon/k_B} = \frac{k_B T}{\varepsilon}. \qquad (6.21)$$

The QuickBASIC program MC2.BAS does a two-dimensional Monte Carlo calculation for 50 Lennard-Jones molecules in a square simulation box and calculates the distribution function $g(r)$.

6.6   Electrical Properties of Solid Surfaces

Solid surfaces are often charged electrically. The surface of a metallic cathode electrode, for example, might lose electrons to a solution component and (with the rest of the electrode) become positively charged. The surface of a colloidal particle might adsorb anions from a solution and become negatively charged. The surface of a crystallite of a slightly soluble salt might lose more anions than cations to the solution and become positively charged.

The *Gouy–Chapman theory* of the electrical double layer at an interface between a solid and a solution describes these processes. The theory calculates the potential ϕ at a distance x from a surface with

$$y = 2 \ln\left(\frac{e^{y_0/2} + 1 + (e^{y_0/2} - 1)e^{-\kappa x}}{e^{y_0/2} + 1 - (e^{y_0/2} - 1)e^{-\kappa x}}\right), \qquad (6.22)$$

where y and κ are defined

$$y = \frac{ze\phi}{k_B T} \qquad (6.23)$$

and

$$\kappa = \left(\frac{2c_\infty z^2 e^2}{\varepsilon_0 \varepsilon_r k_B T}\right)^{1/2}. \qquad (6.24)$$

In the last two statements, z is the charge number for ions in the solution (assumed in our examples to be the same for both anions and cations),

e is the electronic charge, ε_0 is the vacuum permittivity, ε_r is the relative permittivity (dielectric constant), and c_∞ is the molecular concentration of electrolyte in the bulk of the solution. Note that y_0 in Eq. (6.22) is calculated with

$$y_0 = \frac{ze\phi_0}{k_B T},$$

with ϕ_0 the potential on the surface. The charge density σ is calculated with

$$\sigma = (8c_\infty \varepsilon_o \varepsilon_r k_B T)^{1/2} \sinh(y_0/2), \qquad (6.25)$$

and molecular concentrations c_+ and c_- of cations and anions at a distance x from the surface are obtained with

$$c_+ = c_\infty e^{-y} \quad \text{and} \quad c_- = c_\infty e^{y}. \qquad (6.26)$$

Surface charge densities σ are calculated with Eq. (6.25) by the program Gouy1; potentials ϕ at various distances x from the surface with Eqs. (6.22), (6.23), and (6.24) by the program Gouy2; and ionic concentrations at various distances x with Eqs. (6.22), (6.23), (6.24), and (6.26) by the program Gouy3.

6.7   Surface Crystallography

Our concern here is with the surface phase that forms between a crystalline solid phase and a gas phase. This phase is likely to be no more than a few atoms in thickness, and built on an exposed *substrate lattice* determined by the underlying bulk phase. Such a surface can be "clean" meaning that it contains just atoms of the kind that come from the bulk phase, or it can contain foreign atoms adsorbed from the gas phase. A surface phase on a solid is likely to form a two-dimensional *surface lattice* of its own, which may be *commensurate* with the substrate, that is, to some extent it follows the periodicity and orientation of atoms in the substrate lattice, or it may be *incommensurate* and have nothing in common with the substrate lattice.

The key to a commensurate surface lattice is the two-dimensional substrate lattice exposed when the three-dimensional lattice is cleaved along planes with definite Miller indices h, k, l. The formalism of the calculation begins with the two vectors

$$\mathbf{a_1} = (a_{11}, a_{12}) \quad \text{and} \quad \mathbf{a_2} = (a_{21}, a_{22}),$$

which generate the substrate direct lattice. These vectors are combined in the rows of the matrix $\mathbf{a}$,

$$\mathbf{a} = \begin{pmatrix} a_{11} & a_{12} \\ a_{21} & a_{22} \end{pmatrix}. \qquad (6.27)$$

Similarly, the vectors

$$\mathbf{b}_1 = (b_{11}, b_{12}) \quad \text{and} \quad \mathbf{b}_2 = (b_{21}, b_{22})$$

generate the surface direct lattice and are combined in the matrix $\mathbf{b}$,

$$\mathbf{b} = \begin{pmatrix} b_{11} & b_{12} \\ b_{21} & b_{22} \end{pmatrix}. \tag{6.28}$$

A matrix $\mathbf{M}$ connects $\mathbf{a}$ and $\mathbf{b}$,

$$\mathbf{b} = \mathbf{M}\mathbf{a}. \tag{6.29}$$

The matrices $\mathbf{a}^*$ and $\mathbf{b}^*$ contain the vectors $\mathbf{a}_1^*, \mathbf{a}_2^*, \mathbf{b}_1^*$, and $\mathbf{b}_2^*$, which generate the reciprocal substrate and surface lattices. The connection between $\mathbf{a}^*$ and $\mathbf{b}^*$ is

$$\mathbf{b}^* = \mathbf{M}^*\mathbf{a}^*. \tag{6.30}$$

Connections between $\mathbf{a}^*$ and $\mathbf{M}^*$ for the reciprocal lattice and $\mathbf{a}$ and $\mathbf{M}$ for the direct lattice are

$$\mathbf{a}^* = (\tilde{\mathbf{a}})^{-1}, \tag{6.31}$$

and

$$\mathbf{M}^* = (\tilde{\mathbf{M}})^{-1}. \tag{6.32}$$

The surface reciprocal lattice is important because it has the same pattern as that observed with the methods of *low-energy electron diffraction (LEED)*.

The program Leed simulates substrate and surface, direct and reciprocal lattices. It requires data for $\mathbf{a}$ and $\mathbf{M}$ as input, calculates $\mathbf{a}^*$ with Eq. (6.31), $\mathbf{M}^*$ with Eq. (6.32), $\mathbf{b}$ with Eq. (6.29) and $\mathbf{b}^*$ with Eq. (6.30). Table 6.1 defines components of the matrix $\mathbf{a}$ for some common substrate lattices.

TABLE 6.1. Components of the matrix $\mathbf{a}$ for common substrate lattices (*)

Lattice	a_{11}/a	a_{12}/a	a_{21}/a	a_{22}/a
fcc(100)	1/2	1/2	−1/2	1/2
fcc(110)	$\sqrt{2}/2$	0	0	1
fcc(111)	$\sqrt{2}/2$	0	$\sqrt{2}/4$	$\sqrt{6}/4$
bcc(100)	1	0	0	1
bcc(110)	$\sqrt{2}/2$	1/2	$-\sqrt{2}/2$	1/2
bcc(111)	$\sqrt{2}/2$	$\sqrt{6}/6$	$-\sqrt{2}/2$	$\sqrt{6}/6$

*Matrix components are given in reduced form compared to the lattice parameter a.

6.8   Exercises

X-Ray Crystallography

6-1   Use the program `Powder` to assign powder x-ray diffraction data in the file `sio2.dat` to the sets of planes responsible for the reflections.

6-2   Use the Program `Powder` to assign powder x-ray diffraction data in the file `zrsio4.dat` to the sets of planes responsible for the reflections.

6-3   Use the program `Powder` to assign powder x-ray diffraction data in the file `fe2sio4.dat` to the sets of planes responsible for the reflections.

6-4   Modify the program `Xray1` so it plots just the diffraction spot pattern and then run this program for one atom in the unit cell located at $0, 0$ with the choices listed below. How do the cell size and x-ray wavelength affect the number of spots in the diffraction pattern?

Unit cell dimensions/Å	X-ray wavelength/Å
10, 10	5
5, 5	5
10, 10	2.5
5, 5	2.5
20, 20	5

6-5   The x-ray wavelength used to obtain a diffraction pattern is important in determining the resolution of the calculated electron density map. Demonstrate this point by making calculations for four different wavelengths, $\lambda =$ 4, 6, 8, and 10 Å. Use the program `Xray2` and place four atoms in the unit cell at $X_1 = 0.15$, $Y_1 = 0.25$, $X_2 = 0.25$, $Y_2 = 0.65$, $X_3 = 0.80$, $Y_3 = 0.35$, and $X_4 = 0.75$, $Y_4 = 0.75$. Assume $a = 10$ Å and $b = 8$ Å for the unit-cell dimensions.

6-6   Use the program `Xray2` to plot a diffraction spot pattern and calculate an electron density map for an invented triatomic molecule placed anywhere in a unit cell whose dimensions are $a = 8$ Å and $b = 8$ Å. Assume that the x-ray wavelength is $\lambda = 4$ Å. Do the same calculation again for the same molecule with the same orientation, but placed differently in the unit cell. How are the two calculations different and how are they identical?

6-7   Lattices with certain symmetries generate diffraction patterns with systematic absences, that is, certain spots systematically located in the pattern are completely missing. Use an adaptation of the program `Xray2` to display spot patterns for the five cases listed below, all involving a square unit cell with dimensions $a = b = 10$ Å and an x-ray wavelength $\lambda = 5$ Å. In each case, state the rule that governs the presence of observed spots.

	X_1	Y_1	X_2	Y_2	X_3	Y_3	X_4	Y_4
(a)	0.2	0.4	0.8	0.9	–	–	–	–
(b)	0.2	0.4	0.8	0.4	0.7	0.9	0.3	0.9
(c)	0	0.5	0.5	0.5	–	–	–	–
(d)	0.15	0.45	0.85	0.55	0.35	0.95	0.65	0.05
(e)	0.5	0	0	0.5	–	–	–	–

6-8 Use the program `Pattersn` to sketch an idealized Patterson map corresponding to a unit cell with atoms located as follows: $X_1 = 0.1$, $Y_1 = 0.2$, $X_2 = 0.3$, $Y_2 = 0.8$, $X_3 = 0.6$, $Y_3 = 0.3$, and $X_4 = 0.8$, $Y_4 = 0.6$.

6-9 Use an adaptation of the program `Xray4` to plot a Patterson map for atoms located in the unit cell given in the last exercise. Do the calculation first with $\lambda = 4$ Å, then with $\lambda = 1$ Å and 2 Å. Assume that the unit cell dimensions are $a = 10$ Å and $b = 8$ Å. Compare each calculation with the idealized Patterson map sketched in the last exercise.

6-10 Use an adaptation of the program `Xray4` to calculate a Patterson map for a lattice based on a unit cell containing three atoms located at: $X_1 = 0.15$, $Y_1 = 0.25$, $X_2 = 0.75$, $Y_2 = 0.75$, $X_3 = 0.8$, and $Y_3 = 0.35$. Use $a = 10$ Å and $b = 8$ Å for the unit cell dimensions and $\lambda = 2$ Å for the x-ray wavelength. Interpret the result by making an idealized sketch of the Patterson map with the program `Pattersn`.

Electron Diffraction in Gases

6-11 Data are given below for electron diffraction scattering by $SiF_4(g)$. Use these data and an adaptation of the program `Elecdiff` to calculate a value for the bond distance R_{SiF} in the molecule. $SiF_4(g)$ has tetrahedral geometry. Note that the (nonbonding) F-F distance R_{FF} is related to the Si-F (bonding) distance R_{SiF} by $R_{FF} = (8/3)^{1/2} R_{SiF}$.

s_{min}/nm^{-1}	22.19	69.32	115.9	150.8	193.2
s_{max}/nm^{-1}	54.06	82.66	130.2	172.7	213.4

6-12 Data are given below for electron diffraction scattering by $BF_3(g)$. Use these data and an adaptation of the program `Elecdiff` to calculate a value for the bond distance R_{BF} in the molecule. BF_3 has planar-trigonal geometry. Note that the (nonbonding) distance R_{FF} is related to the (bonding) distance R_{BF} by $R_{FF} = \sqrt{3} R_{BF}$.

s_{min}/nm^{-1}	22.9	46.1	79	104
s_{max}/nm^{-1}	34.4	62.2	94	117

6-13 Data are given below for electron diffraction scattering by $CO_2(g)$.

Use these data and an adaptation of the program Elecdiff to calculate a value for the bond distance R_{CO} in the molecule. CO_2 has linear geometry.

s_{min}/nm^{-1}	44	100	154	210
s_{max}/nm^{-1}	67	122	178	230

Semiconductors

6-14 The Fermi level for a pure-material semiconductor is always $E_f = E_g/2$. Use the program Fermi to calculate $[e^-]_c$, $[h^+]_v$ and E_f at 300 K for silicon whose gap energy is $E_g = 1.11$ eV.

6-15 An impurity silicon semiconductor is prepared with $[D]_0 = 10^{14}$ cm^{-3}, $[A]_0 = 10^{15}$ cm^{-3}, $E_v = 0$, $E_g = 1.11$ eV, $E_a = 0.08$ eV, and $E_d = 1.05$ eV. Use the program Fermi to calculate $[e^-]_c$, $[h^+]_v$, and E_f at 300 K. Is this an n-type or p-type semiconductor?

6-16 At low temperatures in an n-type semiconductor most of the electron carriers are created by ionizations of donor levels, the Fermi level is located near the top of the energy gap, and $[e^-]_c \gg [h^+]_v$. But as the temperature is raised more electron carriers are created by ionizations from the valence band and the Fermi level is lowered. At high temperatures, the Fermi level approaches the middle of the energy gap and $[e^-]_c \cong [h^+]_v$, that is, the semiconductor behaves as if it had no impurities. Demonstrate these trends by running an adaptation of the program Fermi with $[D]_0 = 10^{14}$ cm^{-3}, $[A]_0 = 0$, $E_v = 0$, $E_g = 1.11$ eV (silicon), $E_a = 0.08$ eV, and $E_d = 1.05$ eV, and for temperatures in the range 200–700 K. Revise Fermi so it makes the calculation in a loop covering the required temperature range.

Molecular Dynamics and Monte Carlo Simulations

The following three exercises make use of the QuickBASIC programs MC1.BAS, MC2.BAS, MD1.BAS, MD2.BAS, and MD3.BAS. All of these programs have graphics features requiring data from the file GRAPH.DAT. To create this file run the QuickBASIC program PIXEL.BAS and enter the graphics mode you are using (choices are CGA, EGA, and VGA). Three of the programs, MC1.BAS, MC2.BAS, and MD3.BAS, also require the subprogram PLOTAXES.BAS. To run one of these programs, MC1.BAS for example, *open* MC1.BAS first and then *load* PLOTAXES.BAS.

6-17 Molecular dynamics models simulate the melting transition. When the molecules are closely packed in a lattice, their movements are confined to small "cages" surrounding the lattice positions. As the packing is made looser (by increasing the ratio V/V_0), the molecular motion becomes correspondingly less confined until some of the molecules are able to exchange places with molecules in neighboring cages. Caged molecules simulate behavior of the solid phase, and *molecules* moving freely between cages simulate the liquid phase. The molecular dynamics simulation supplied by the program MD1.BAS finds a sharp transition between solid and liquid behavior when V/V_0 has values in the range 1.58 to 1.60. Demonstrate this by running MD1.BAS for 3000 collisions and with values of V/V_0 in the range indicated.

6-18 The program MD2.BAS displays in an animation the motion of one molecule as it bumps, and is bumped by, its neighbors. Run the program with $V/V_0 = 2$ and for 1000 collisions. Watch the progress of the emphasized molecule. If it moves beyond the edge of the square simulation box it is replaced by another molecule on the opposite side of the box.

6-19 The programs MC1.BAS and MC2.BAS (and MD3.BAS) calculate and plot the radial distribution function $g(r)$ for simulated liquid structures. This function is defined so that if ρ is the bulk density in the liquid

$$\rho g(r) = \text{average density of molecules at a radial distance } r$$
$$\text{from a particular molecule.}$$

Run MC1.BAS for $V/V_0 = 1.5$ and 1000 calculation cycles, and note that $g(r)$ has several maxima—a large one for $r = \sigma$ (σ is a molecular diameter) and smaller ones for $r = 2\sigma$ and 3σ. The maxima locate *coordination shells* of molecules surrounding the particular molecule. There is one well-defined shell at $r = \sigma$, and smaller ones for $r = 2\sigma$ and 3σ. Also run the program MC2.BAS for $V/V_0 = 1.5$, 1000 calculation cycles and $T^* = 1$. Compare with the run of MC1.BAS. Account for the similarities and differences.

Electrical Properties of Solid Surfaces

6-20 A charged surface is in contact with an aqueous 0.0100 mol L^{-1} solution of NaCl. Each atomic site on the surface occupies 2.0 nm^2, and one in four sites accomodates a chloride ion; sodium ions are not adsorbed on the surface. Write a program that uses the Gouy–Chapman theory to calculate the potential $\phi(x)$ for $x = 2.0$ nm. The temperature is 298 K. Also have the program calculate the potential ϕ_0 on the surface.

Surface Crystallography

6-21 A face-centered cubic lattice is cleaved along the (110) planes and a (2×1) surface lattice forms on this substrate lattice. For this surface lattice,

$$\mathbf{M} = \begin{pmatrix} 2 & 0 \\ 0 & 1 \end{pmatrix}.$$

Use the program `Leed` to plot substrate and surface, direct and inverse lattices for this case.

6-22 A face-centered cubic lattice is cleaved along the (100) planes and a $c(2 \times 2)$ surface lattice forms on this substrate lattice. For this surface lattice,

$$\mathbf{M} = \begin{pmatrix} 1 & -1 \\ 1 & 1 \end{pmatrix}.$$

Use the program `Leed` to plot substrate and surface, direct and inverse lattices for this case.

6-23 A body-centered cubic lattice is cleaved along the (100) planes and a (2×2) surface lattice forms on this substrate lattice. For this surface lattice,

$$\mathbf{M} = \begin{pmatrix} 2 & 0 \\ 0 & 2 \end{pmatrix}.$$

Use the program `Leed` to plot substrate and surface, direct and inverse lattices for this case.

7
Macromolecules

Macromolecules are marvels of complexity and diversity, as we all know. To a large extent we *are* macromolecules. A detailed study of the structure of a macromolecule (e.g., with x-ray diffraction methods) is an arduous task, requiring much skill, patience, and ingenuity. But if you are willing to forgo some of the details and picture a macromolecule as a random coil, a light-scattering particle, or a hydrodynamic particle with an ellipsoid shape, the analysis is much simpler—in fact, almost easy. This chapter emphasizes these almost-easy approaches to macromolecule structure. In Sec. 7.1 the random coil model is used and dimensions are calculated with statistical formulas (the programs `Coil1`, `Coil2`, and `Coil3`). In Sec. 7.2 the ellipsoid model is introduced and applied to determination of a macromolecule's shape and dimensions (the programs `Perrin1` and `Perrin2`). This calculation is based on measurements of diffusion coefficients. Another approach, beginning with viscometry data, also relies on the ellipsoid model and leads to approximately the same results (the programs `Simha1` and `Simha2`). A third method presented in Sec. 7.4, pictures a macromolecule as a light-scattering particle. The light-scattering measurements also supply data on a macromolecule's size and shape (the program `Zimm`).

7.1 Random Coils

Some macromolecules have an extended structure with the segments moving freely with respect to each other, "writhing and twisting and changing shape," as Richards puts it; such a molecule has no definite shape. At another extreme are macromolecules that fold themselves into tight, highly organized structures. We first discuss the loose structures, called *random coils*.

Picture the random coil as a sequence of linked bonds joining monomer residues in the chain. Assume that there are n residues with $n - 1$ bonds joining them, and that each bond has the length L. At each residue, for example R_i, two angles θ_i and ϕ_i define the *conformation*, or three-dimensional structure, of the molecule (Fig. 7.1). In the simplest model all of the angles θ_i

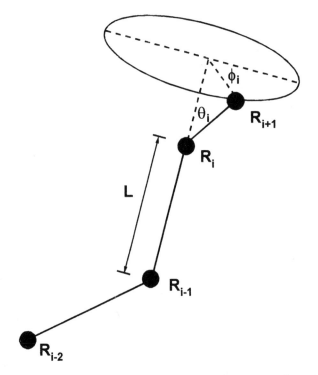

FIGURE 7.1. Part of a random-coil macromolecule. Residues are indicated by circles. Note that the two angles θ_i and ϕ_i define the conformation at the residue R_i.

and ϕ_i are assumed to have random values. This *freely jointed chain* is not a model for any real macromolecule, but it serves as a beginning point for more realistic models.

The program Coil1 plots bond segments for a freely-jointed chain (and for another model considered later). The plot locates three important points on the chain: the two ends and the center of mass. One measure of size of the chain is the end-to-end distance r. Because a freely-jointed chain is random it can have many different conformations and end-to-end distances r, but $\overline{r^2}$, the average of the squared end-to-end distance calculated for many freely-jointed chains, depends in a simple way on the number of bonds $n - 1$, and the bond length L,

$$\overline{r^2} = (n - 1)L^2 \quad \text{(many random conformations)}. \tag{7.1}$$

Another important measure of chain size, the square of the *radius of gyration* R_G^2, is calculated by summing the squares of the distances R_{Gi} from each residue to the chain's center of mass and averaging by dividing by the number of residues,

$$R_G^2 = \frac{\sum\limits_{i=1}^{n} R_{Gi}^2}{n}. \tag{7.2}$$

The radius of gyration, like the end-to-end distance, changes with the conformation of the chain, but an average $\overline{R_G^2}$ of R_G^2 is simply related to $\overline{r^2}$,

$$\overline{R_G^2} = \frac{\overline{r^2}}{6} \quad (n \text{ large}), \tag{7.3}$$

for large chains, so

$$\overline{R_G^2} = \frac{(n-1)L^2}{6} \quad (n \text{ large}). \tag{7.4}$$

The freely-jointed chain is of little value as a model of any real macromolecule, but it is improved by simple modifications. The most obvious failing of the freely-jointed chain is that it allows random values for the angles θ_i. In a simple chain, such as that of polymethylene, the bond angles, that is, the angles between adjacent bonds ($180 - \theta_i$ for residue R_i in Fig. 7.1), are all expected to be approximately the same. A useful model, called the freely rotated chain, assumes a single value for the θ_i's, but random values for the ϕ_i's. In this model, an average of the squared end-to-end distance $\overline{r^2}$ is calculated for many large freely rotated chains according to

$$\overline{r^2} = (n-1)L^2 \frac{1 + \cos\theta}{1 - \cos\theta} \quad (n \text{ large}), \tag{7.5}$$

where θ is the constant value of the θ_i's. The angle between adjacent bonds might typically be the tetrahedral angle 109°, so $\theta = 180 - 109 = 71°$, $\cos\theta > 0$, and according to Eq. (7.5), $\overline{r^2}$ is larger for this more constrained chain.

If the "large n" condition for Eq. (7.5) is not met, a more elaborate equation must be used,

$$\overline{r^2} = C(n-1)L^2, \tag{7.6}$$

with

$$C = \frac{1 + \cos\theta}{1 - \cos\theta} - \frac{2\cos\theta}{n-1}(1 - \cos^{n-1}\theta)(1 - \cos\theta)^{-2}. \tag{7.7}$$

The program Coil1 calculates and plots in three dimensions random chains of both the freely rotated and freely jointed kinds. To run the program enter data beginning on the third line of code for: the bond length L, the mass m of each residue, the number of residues n, and

```
chain = "fr";
```

if you want a freely rotated chain, or

```
chain = "fj";
```

for a freely jointed chain. If the chain is freely rotated, then also enter a value for the bond angle (in degrees), for example,

```
bondAngle = 135;
```

The program plots the chain in three dimensions and also calculates values for the radius of gyration R_G and the end-to-end distance r.

The program `Coil1` calculates r directly from the program data for locations of the first and nth residues. The programs `Coil2` and `Coil3` make the same calculation and then compare with Eq. (7.6) for a freely rotated chain (in `Coil2`), or with Eq. (7.1) for a freely jointed chain (in `Coil3`).

7.2   Macromolecules as Hydrodynamic Particles

Imagine that a dissolved macromolecule A is disturbed by an applied centrifugal, electrical, or diffusional force. The macromolecular particle accelerates at first, but at the same time it is affected by an increasing opposing frictional force as a result of physical interactions between the particle and the solvent medium through which it moves. A dynamic steady-state condition is rapidly reached in which the applied and frictional forces are balanced, and the particle attains a constant *terminal speed* w_A. If the applied force is F_A per mole of component A, the average force per molecule is F_A/L ($L =$ Avogadro's constant), and we note that the terminal speed w_A increases in direct proportion to this molecular force,

$$f_A w_A = \frac{F_A}{L}, \qquad (7.8)$$

where f_A, called a *frictional coefficient*, is a much-used parameter in the study of macromolecule behavior. Frictional coefficients are measurable, as we will see, and they depend on the size and shape of the molecule, and also on the viscosity of the solvent medium.

The viscosity dependence is brought out in *Stokes Law*, which is valid for molecules with spherical shapes,

$$f_A = 6\pi\eta r_{0A} \quad \text{(spherical molecules)}, \qquad (7.9)$$

in which η is the viscosity coefficient of the medium and r_{0A} is the radius of the spherical molecules.

The volume of a macromolecule A with molar mass M_A and partial specific volume v_A is $M_A v_A/L$. If the macromolecule has incorporated solvent molecules (e.g., water in proteins) the volume V_s of this solvated molecule is

$$V_s = \frac{M_A(v_A + \delta_B v_B^*)}{L}, \qquad (7.10)$$

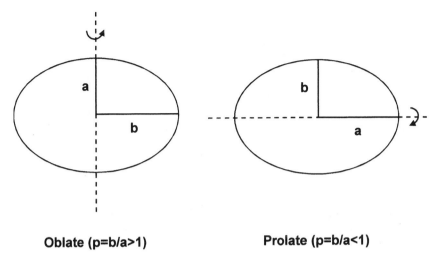

Oblate (p=b/a>1)   **Prolate (p=b/a<1)**

FIGURE 7.2. Prolate and oblate ellipsoids of revolution seen in cross section. Three-dimensional shapes are generated by rotating around the axes indicated.

where B represents the solvent, δ_B is the mass of solvent molecules per gram of macromolecule, and v_B^* is the specific volume of the pure solvent. For a spherical molecule

$$r_{0A} = \left(\frac{3M_A(v_A + \delta_B v_B^*)}{4\pi L}\right)^{1/3}. \tag{7.11}$$

Calculation of the frictional coefficient f_A becomes considerably more complicated when the macromolecules do not have spherical shapes. The detailed calculation was done by Perrin for two idealized distortions of the spherical shape called *prolate* and *oblate ellipsoids of revolution* (Fig. 7.2). These figures are defined by the dimensions a and b measured along the axis of revolution and along a perpendicular (equatorial) axis. The *axial ratio* $p = b/a$ is important in Perrin's analysis: note that $p < 1$ for prolate shapes, $p > 1$ for oblate shapes and $p = 1$ for a sphere. The volume for any of these figures is

$$V = \frac{4\pi ab^2}{3}. \tag{7.12}$$

Perrin's calculation determines the *frictional ratio* f_{rA} for a macromolecule A, defined as

$$f_{rA} = \frac{f_A}{f_{0A}}, \tag{7.13}$$

where $f_{0A} = 6\pi\eta r_{0A}$ is the frictional coefficient the macromolecule would have if it were spherical with radius r_{0A} [Eq. (7.9)]. The frictional ratio is important because it depends only on the shape of the molecule and not on its size.

The program `Perrin1` performs the Perrin calculation and plots the frictional ratio f_r against the axial ratio $p = b/a$. Consult the program for details on the equations involved. Note that the minimum value of the frictional ratio is $f_r = 1$ for a spherical shape.

The program `Perrin2` does a specific Perrin calculation, either the frictional ratio f_r from an axial ratio p, or vice versa, p from f_r. The next example illustrates.

Example 7-1. A macromolecule has the volume 8380 nm^3, and it is known to have an approximately prolate shape. Its frictional coefficient is $f_r = 1.044$. Calculate the approximate dimensions of the molecule.

Answer. Enter the frictional ratio in the program `Perrin2`,

$$fr = 1.044;$$

and the prolate shape,

$$shape = "prolate";$$

and calculate the axial ratio $p = 0.50$. Thus $a = b/0.50$, and from the given volume and Eq. (7.12),

$$8380 \text{ nm}^3 = \frac{4\pi a b^2}{3} = \frac{4\pi b^3}{(3)(0.50)},$$

so

$$b = \left(\frac{(3)(8380 \text{ nm}^3)(0.50)}{4\pi}\right)^{1/3},$$

$$= 10 \text{ nm},$$

and

$$a = \frac{b}{0.50} = \frac{10 \text{ nm}}{0.50}$$

$$= 20 \text{ nm}.$$

7.3   Diffusion

In the last section we found methods for calculating the frictional ratio of a macromolecule from information on the size and shape of the molecule. Here we determine how to obtain an *experimental* value for the frictional ratio, and then from this, through the program `Perrin2`, the molecule's size and shape.

The key experimental measurement is that of diffusion. Molecules of any kind diffuse under the influence of a diffusional force. For a diffusing component A this force, call it F_{dA}, in a direction x is calculated as the negative gradient of the chemical potential μ_A,

$$F_{dA} = -\frac{d\mu_A}{dx}. \tag{7.14}$$

In systems dilute enough to guarantee ideal solution behavior $\mu_A = \mu_A^0 + RT \ln C_A$, where C_A is the molar concentration of A (Sec. 2.6), and

$$F_{dA} = -RT \frac{d \ln C_A}{dx}.$$

This is one of the molar forces that can be used in Eq. (7.8) containing the frictional coefficient f_A. We find that

$$f_A w_A = \frac{F_{dA}}{L} = -\frac{RT}{L} \frac{d \ln C_A}{dx},$$

where w_A is now the average speed of A molecules moving in the x direction. Noting that $d \ln C_A = dC_A/C_A$, we rearrange the equation to

$$w_A C_A = -\frac{RT}{f_A L} \frac{dC_A}{dx}. \tag{7.15}$$

The quantity on the left side of Eq. (7.15) expresses the *molar flux* J_A of A molecules, that is, the number of A molecules passing through a plane perpendicular to x per unit area per second,

$$J_A = w_A C_A.$$

Thus, Eq. (7.15) can be written

$$J_A = -\frac{RT}{f_A L} \frac{dC_A}{dx}. \tag{7.16}$$

This equation has the same form as one of the basic empirical laws of diffusion,

$$J_A = -D_A \frac{dC_A}{dx}, \tag{7.17}$$

in which D_A is a measurable parameter called the diffusion coefficient. With C_A expressed in mol cm^{-3}, D_A has the units commonly used, cm^{-2} s^{-1}.

Comparison of Eqs. (7.16) and (7.17) brings us to an equation that permits calculation of a frictional coefficient f_A from a measured diffusion coefficient D_A,

$$f_A = \frac{RT}{D_A L}. \tag{7.18}$$

Since it has been derived for ideal solutions this equation usually applies only to very dilute solutions. Diffusion coefficients used in the equation are customarily obtained by extrapolation to the condition of infinite dilution. The use of Eq. (7.18) is demonstrated in the next example.

Example 7-2. The protein myosin has the molar mass 4.93×10^5 g mol^{-1}. In very dilute aqueous solutions, its diffusion coefficient is $D_A = 1.10 \times 10^{-7}$ cm^{-2} s^{-1} at 20.0 °C. Calculate the frictional coefficient f_A for myosin. Also calculate the frictional ratio $f_{rA} = f_A/f_{0A}$, assuming that $v_A = 0.728$ cm^3 g^{-1} and that the macromolecule has no incorporated solvent [$\delta_B = 0$ in Eq. (7.11)]. Finally, calculate the axial ratio $p = b/a$ and the molecule's dimensions a and b, assuming a prolate shape. Water has the viscosity coefficient $\eta = 1.002 \times 10^{-3}$ kg m^{-1} s^{-1} at 20.0 °C.

Answer. Calculate f_A with Eq. (7.18)

$$f_A = \frac{(8.3145 \text{ J mol}^{-1} \text{ K}^{-1})(293.0 \text{ K})}{(1.10 \times 10^{-7} \text{ cm}^2 \text{ s}^{-1})(10^{-4} \text{ m}^2 \text{ cm}^{-2})(6.0221 \times 10^{23} \text{ mol}^{-1})}$$

$$= 3.68 \times 10^{-10} \text{ kg s}^{-1}.$$

Calculate r_{0A}, the radius the molecule would have if it were spherical, using Eq. (7.11) and $\delta_B = 0$,

$$r_{0A} = \left(\frac{(3)(4.93 \times 10^5 \text{ g mol}^{-1})(0.728 \text{ cm}^3 \text{ g}^{-1})}{(4\pi)(6.0221 \times 10^{23} \text{ mol}^{-1})} \right)^{1/3}$$

$$= 5.221 \times 10^{-7} \text{ cm}$$

$$= 5.221 \text{ nm}.$$

Calculate f_{0A} using Eq. (7.9) and $\eta = 1.002 \times 10^{-3}$ kg m^{-1} s^{-1} for water,

$$f_{0A} = (6\pi)(1.002 \times 10^{-3} \text{ kg m}^{-1} \text{ s}^{-1})(5.221 \times 10^{-9} \text{ m})$$

$$= 9.86 \times 10^{-11} \text{ kg s}^{-1}.$$

Then, the frictional ratio is

$$f_{rA} = \frac{f_A}{f_{0A}} = \frac{3.68 \times 10^{-10} \text{ kg s}^{-1}}{9.86 \times 10^{-11} \text{ kg s}^{-1}}$$

$$= 3.73,$$

and from this the program Perrin2 calculates the axial ratio $p = a/b = 0.0120$. Calculate the dimensions a and b separately by first obtaining the molecule's volume from Eq. (7.10) with $\delta_B = 0$,

$$V = \frac{M_A v_A}{L}$$

$$= \frac{(4.93 \times 10^5 \text{ g mol}^{-1})(0.728 \text{ cm}^3 \text{ g}^{-1})}{(6.0221 \times 10^{23} \text{ mol}^{-1})}$$

$$= 5.96 \times 10^{-19} \text{ cm}^3$$

$$= 5.96 \times 10^{-25} \text{ m}^3.$$

Then combine Eq. (7.12), with $a = b/(0.0120)$ obtained above and the volume just calculated to arrive at values of a and b,

$$b = \left(\frac{(3)(5.96 \times 10^{-25} \text{ m}^3)(0.0120)}{4\pi}\right)^{1/3}$$

$$= 1.20 \times 10^{-9} \text{ m}$$

$$= 1.20 \text{ nm}$$

$$a = \frac{b}{0.0120} = \frac{1.20 \text{ nm}}{0.0120} = 100 \text{ nm}.$$

Myosin is a muscle protein; it has an elongated fiber-like shape, as these dimensions indicate.

7.4   Viscometry

Viscometry is utilized in the study of macromolecules by making viscosity measurements on a series of dilute solutions of the macromolecule. The key measured quantity is the *relative viscosity* $\eta_r = \eta_s/\eta_0$, which compares the viscosity coefficient η_s of a solution with the viscosity coefficient η_0 of the pure solvent. The relative viscosity η_r has a virial-like dependence on the mass concentration g_A (units g L^{-1}) of the macromolecule in the solution,

$$\eta_r = 1 + [\eta]_A g_A + C g_A^2 + \dots, \tag{7.19}$$

where $[\eta]_A$, called the *intrinsic viscosity* of component A (although it does not have viscosity units), and C are constants.

The importance of the intrinsic viscosity $[\eta]_A$ is that it depends on the molecular volume of the dissolved macromolecule, and also, in a sensitive way, on molecular shape. The following equation expresses these dependencies for a solvated macromolecule whose volume is V_s,

$$[\eta]_A = \frac{v_A V_s L}{M_A}, \tag{7.20}$$

in which v_A is an important shape-dependent parameter introduced by Simha and called the *Simha factor*. From Eq. (7.10), we see that Eq. (7.20) can be expressed more conveniently

$$[\eta]_A = v_A(v_A + \delta_B v_B^*). \tag{7.21}$$

The Simha factor v_A, like the frictional factor f_{rA}, depends only on molecular shape, and that dependence can be expressed as a function of the axial ratio $p = b/a$ for prolate and oblate ellipsoids of revolution. The program Simha1 does the calculations and plots v_A vs p for both cases. Values of v_A, and of the intrinsic viscosity $[\eta]_A$, become very large for elongated prolate molecules. The minimum value for the Simha factor is $v_A = 2.5$ for a spherical shape. The program Simha2 does a specific Simha calculation, either the Simha factor v_A from the axial ratio p, or vice versa, p from v_A. Here is an example showing how the Simha factor is used.

Example 7-3. Myosin has the molar mass $M_A = 4.93 \times 10^5$ g mol^{-1}, intrinsic viscosity $[\eta]_A = 217$ cm^3 g^{-1} in aqueous solutions at 20.0 °C, and the partial specific volume $v_A = 0.728$ cm^3 g^{-1}. Calculate the Simha factor v_A and the axial ratio p for the molecule, assuming that it has no bound solvent ($\delta_B = 0$) and a prolate shape.

Answer. Calculate v_A using Eq. (7.21) and $\delta_B = 0$,

$$v_A = \frac{[\eta]_A}{v_A} = \frac{217 \text{ cm}^3 \text{ g}^{-1}}{0.728 \text{ cm}^3 \text{ g}^{-1}}$$

$$= 298.$$

Enter this result in the program Simha2, assume a prolate shape and calculate $p = b/a = 0.0148$. This agrees about as well as can be expected with the axial ratio ($b/a = 0.0120$) calculated for myosin from diffusion coefficient data in Example 7-2.

7.5   Macromolecules as Light-Scattering Particles

Macromolecules in solution scatter light away from the direction of an incident beam. Scattering can change the energies of the incident photons, as in Raman spectroscopy, or it can leave photon energies unaffected. Our concern here is with the latter, called *Rayleigh scattering*. Such scattering is measured in an instrument that gathers light scattered at an angle θ from a solution of the macromolecule with volume V located at a distance r_S from the detector. The ratio I_S/I_0 of scattered intensity I_S to the incident intensity

I_0 is directly proportional to V, inversely proportional to r_S^2, and scattering at low angles is greater than that at larger angles, as expressed by the function $1 + \cos^2 \theta$,

$$\frac{I_S}{I_0} \propto \frac{V(1 + \cos^2 \theta)}{r_S^2},$$

or

$$\frac{I_S}{I_0} = \frac{R_\theta V(1 + \cos^2 \theta)}{r_S^2}, \tag{7.22}$$

in which R_θ is a constant called the *Rayleigh ratio*.

Light-scattering measurements are useful in the study of macromolecules because the extent of the scattering depends on the size and shape of the macromolecules responsible for the scattering. If the macromolecules are small compared to the wavelength of the incident light (although still large enough to be called macromolecules), if the solution containing the molecules is ideal, and if the solution is monodisperse, the Rayleigh ratio R_θ depends just on the molar mass M_A of the macromolecules and their mass concentration g_A in the solution,

$R_\theta = K M_A g_A$ (small macromolecules; ideal, monodisperse solutions). (7.23)

The factor K is a constant containing all the optical parameters that characterize the scattering experiment.

In light-scattering experiments involving macromolecules, nonideality is the rule. Thus Eq. (7.23) requires modification for application to real solutions. The equation most often used with nonideal solutions of small macromolecules has the virial form,

$$\frac{K g_A}{R_\theta} = \frac{1}{M_A} + 2 A_2 g_A \quad \begin{array}{l} \text{(small macromolecules; nonideal,} \\ \text{monodisperse solutions),} \end{array} \tag{7.24}$$

in which g_A is again the mass concentration of macromolecules in the solution, and A_2 is a virial coefficient. In an ideal solution $A_2 = 0$ and Eq. (7.24) reduces to (7.23).

If the macromolecules are not small compared to the wavelength of the incident light, Eq. (7.24) has to be further modified by including a shape factor which depends on the scattering angle θ. We write this factor $P(\theta)$ and note that it is included only in the first term of Eq. (7.24),

$$\frac{K g_A}{R_\theta} = \frac{1}{P(\theta) M_A} + 2 A_2 g_A \quad \begin{array}{l} \text{(large macromolecules; nonideal,} \\ \text{monodisperse solutions).} \end{array} \tag{7.25}$$

Since Eq. (7.25) reduces to Eq. (7.24) for small molecules, $P(\theta)$ must approach one for small-sized macromolecules. For large macromolecules the

$P(\theta)$ factor depends on the square of the macromolecule's radius of gyration according to the approximate equation,

$$\frac{1}{P(\theta)} \cong 1 + \frac{16\pi^2 \overline{R_G^2}}{3\lambda^2} \sin^2(\theta/2). \tag{7.26}$$

Light-scattering measurements are usually aimed at determining the molar mass of the macromolecule studied. Eq. (7.25) provides that information in a double extrapolation of θ and g_A to zero values, so $P(\theta) = 1$, the virial term containing A_2 in Eq. (7.25) vanishes, and $Kg_A/R_\theta = 1/M_A$, as in the ideal equation (7.23). A procedure developed by Zimm accomplishes this extrapolation. Zimm's technique is to plot Kg_A/R_θ, call this function y,

$$y = \frac{Kg_A}{R_\theta}, \tag{7.27}$$

against a function that depends on both θ and g_A, call it x,

$$x = \sin^2(\theta/2) + cg_A. \tag{7.28}$$

The constant c preceding g_A is chosen to make the plot convenient but is otherwise arbitrary.

The program Zimm simulates a typical Zimm plot. Input to the program consists of value for the light wavelength lambda (in nm); the radius of gyration RG (in nm); the virial coefficient A2 (in mol cm^3 g^{-2}); the molar mass MA (in g mol^{-1}); and the constant c (in cm^3 g^{-1}). The program also requires a list theta of scattering angles (in degrees) and a list gA of concentrations of solutions (in g cm^{-3}). Data supplied to Zimm are for solutions of cellulose nitrate in acetone at 25 °C.

Run the program and note that the Zimm plot consists of a network of data points (blue) which define constant-θ curves (red) and constant-g_A curves (yellow). Two extrapolated curves are located on the plot (by the green points) and they meet at an intercept where the double extrapolation is in effect and $y = \dfrac{kg_A}{R_\theta} = \dfrac{1}{M_A}$. Slopes of the two extrapolated curves are

$$\left(\frac{\partial y}{\partial x}\right)_{\theta=0} = \frac{2A_2}{c}, \tag{7.29}$$

and

$$\left(\frac{\partial y}{\partial x}\right)_{g_A=0} = \left(\frac{16\pi^2 \overline{R_G^2}}{3\lambda^2}\right)\frac{1}{M_A}. \tag{7.30}$$

Thus, measured slopes of the extrapolated curves can be used to calculate the virial coefficient A_2 and the radius of gyration $(\overline{R_G^2})^{1/2}$ for the macromolecule studied.

Equations (7.23) to (7.30) require the assumption that the solution of macromolecules is monodisperse, so only a single molar mass M_A is present.

Light-scattering data from polydisperse solutions obey equations of the same form with M_A replaced by the mass-average molar mass $\overline{M}_m$, so Eq. (7.25) becomes

$$\frac{Kg_A}{R_\theta} = \frac{1}{P(\theta)\overline{M}_m} + 2A_2g_A \qquad \begin{array}{l} \text{(large macromolecules; nonideal,} \\ \text{polydisperse solutions),} \end{array} \qquad (7.31)$$

and Zimm's double extrapolation provides an intercept, which is equal to $1/\overline{M}_m$.

7.6 Exercises

Random Coils

7-1   Run the program Coil1 for a freely rotated chain with $L = 10, n = 100$ and with the bond angle set at $0°$ and then at $180°$. Account for the shapes of the plots and the calculated end-to-end distances. Do the same thing for $n = 101$. The radius of gyration for these cases is considered in the next exercise.

7-2   Repeat the calculation requested in the last exercise and account for the calculated values of the radius of gyration.

7-3   Revise the program Coil1 so it plots chain structures for macromolecules which have both of the angles θ and ϕ defined in Fig. 7.1 fixed. Run the revised program with $\theta = \phi = 30°$, $n = 101$ and $L = 10$. Compare with the chain generated by $\theta = 30°$ and random values of ϕ.

7-4   Run the program Coil2 with $L = 10$, $n = 17$ and a bond angle of $30°$, and note the comparison of the average end-to-end distance calculated with data from the program and with the statistical formula.

7-5   Run the program Coil3 with $L = 10$, $n = 17$, and note the comparison of the average end-to-end distance calculated with data from the program and with the statistical formula.

Macromolecules as Hydrodynamic Particles

7-6   Molecules of a polystyrene preparation dissolved in benzene at $25.0\,°C$ have the frictional ratio $f_{rA} = 1.548$. The number-average molar mass of the polystyrene is $\overline{M}_n = 1.32 \times 10^3$ g mol^{-1} and the partial specific volume is $v_A = 0.90$ cm^3 g^{-1}. Estimate the dimensions a and b of the molecule modeled as a prolate ellipsoid of revolution. Assume that the molecules have no incorporated solvent.

7-7   Molecules of another polystyrene preparation dissolved in benzene at $25\,°C$ have the frictional ratio $f_{rA} = 3.70$. The number-average molar mass of this polystyrene is $\overline{M}_n = 1.20 \times 10^6$ g mol^{-1} and the partial specific volume is about $v_A = 0.90 \text{ cm}^3 \text{ g}^{-1}$. Estimate the molecular dimensions a and b, assuming that the molecules have prolate shapes and that they have no incorporated solvent.

Diffusion

7-8 A polyvinyl alcohol has the diffusion coefficient $3.97 \times 10^{-7} \text{ cm}^2 \text{ s}^{-1}$ in water at $25.0\,^\circ\text{C}$ in very dilute solutions. The molar mass of this preparation is $2.33 \times 10^4 \text{ g mol}^{-1}$, and its partial specific volume is $0.750 \text{ cm}^3 \text{ g}^{-1}$. Calculate the axial ratio $p = b/a$ for the molecule modeled as a prolate ellipsoid of revolution. Assume that the molecules have no incorporated solvent and that the solutions have the same viscosity coefficient as pure water, that is, $1.002 \times 10^{-3} \text{ kg m}^{-1} \text{ s}^{-1}$.

7-9 A cellulose nitrate preparation has the diffusion coefficient $6.5 \times 10^{-7} \text{ cm}^2 \text{ s}^{-1}$ in acetone at $20.0\,^\circ\text{C}$ in very dilute solutions. The molar mass of this preparation is $1.00 \times 10^5 \text{ g mol}^{-1}$, and its partial specific volume is $0.51 \text{ cm}^3 \text{ g}^{-1}$. Calculate the dimensions a and b for the molecule modeled as a prolate ellipsoid of revolution. Assume that the molecules have no incorporated solvent and that the solutions have the same viscosity coefficient as pure acetone, that is, $3.26 \times 10^{-4} \text{ kg m}^{-1} \text{ s}^{-1}$.

7-10 A preparation of polyvinylpyrrolidine in water at $20.0\,^\circ\text{C}$ has the diffusion coefficient $7.55 \times 10^{-7} \text{ cm}^2 \text{ s}^{-1}$ in very dilute solutions, the axial ratio $p = b/a = 0.076$, and the partial specific volume $0.802 \text{ cm}^3 \text{ g}^{-1}$. Calculate the frictional ratio f_r and then the molar mass of the macromolecules, assuming that they have no incorporated solvent.

Viscometry

7-11 You have measured $[\eta]_A = 820 \text{ cm}^3 \text{ g}^{-1}$ for the intrinsic viscosity of a macromolecule in water, $M_A = 3.2 \times 10^4 \text{ g mol}^{-1}$ for the molar mass, and $v_A = 0.75 \text{ cm}^3 \text{ g}^{-1}$ for the partial specific volume. Use the program Simha2 to calculate the axial ratio $p = b/a$ and the dimensions a and b, assuming a prolate shape and that the molecules contain no incorporated solvent.

7-12 Relative viscosities η_r are measured with a remarkably simple instrument called a *viscometer*. The *efflux time* t_s for a certain volume of the solution of interest to flow through a glass capillary in the viscometer is measured and compared to the efflux time t_0 for the same volume of solvent. An approximate calculation of η_r is then obtained as the ratio t_s/t_0 of t_s to t_0. Efflux times for solutions of a polyvinyl alcohol dissolved in water at $25.0\,^\circ\text{C}$ are quoted below. Use these data, the program Linreg, and Eq. (7.19) rearranged to

$$\frac{\eta_r - 1}{g_A} = [\eta]_A + C g_A \tag{7.32}$$

to calculate a value for the intrinsic viscosity $[\eta]_A$. Intrinsic viscosities are related to an average molar mass $\bar{M}_{vA}$, called the *viscosity-average molar mass*, according to an empirical equation,

$$[\eta]_A = K[\bar{M}_{vA}/(g\,mol^{-1})]^{a_A}, \tag{7.33}$$

where $K = 2.0 \times 10^{-2}\ cm^3\,g^{-1}$ and $a_A = 0.76$ for the aqueous polyvinyl alcohol solutions. Calculate the viscosity-average molar mass for the polyvinyl alcohol.

$g_A/(g\,L^{-1})$	t_s/s
.00	262.30
2.00	285.40
4.00	312.50
6.00	343.70
8.00	378.80
10.00	419.00

7-13 The viscometer efflux times (see Exercise 7-12) listed below are obtained with solutions of an atactic polystyrene in benzene at 25.0 °C. The partial specific volume and molar mass for this polystyrene are 0.90 $cm^3\,g^{-1}$ and 5.6 $\times$ 10^3 $g\,mol^{-1}$. Estimate the dimensions a and b of the molecule, assuming that it has a prolate shape and that it has no incorporated solvent.

$g_A/(g\,L^{-1})$	t_s/s
.00	210.20
2.00	215.50
4.00	227.00
6.00	245.70
8.00	270.70
10.00	302.70

7-14 You are going to do a viscometry study of a macromolecule preparation with no computer at hand. You will need tables of Simha factors calculated as functions of the axial ratio a/b (not $p = b/a$), and vice versa. Revise the program Simha2 so that for prolate ellipsoids of revolution it calculates: (a) v_A for a/b ranging from 1 to 10 in steps of 1; (b) v_A for $a/b = 10$ to 200 in steps of 10; (c) a/b for $v_A = 2.5$ to 10 in steps of 0.5; (d) a/b for $v_A = 10$ to 1000 in steps of 10.

Macromolecules as Light-Scattering Particles

7-15 Use the program Zimm to construct a Zimm plot for a polystyrene preparation in butanone. Relevant parameters for the plot are

$\lambda = 436$ nm, $\bar{R}_G^{1/2} = 46$ nm, $A_2 = 1.29 \times 10^{-4}$ cm^3 mol g^{-2}, $c = 100$ cm^3 g^{-1}, $\bar{M}_m = 1.03 \times 10^6$ g mol^{-1}. Concentrations of the solutions involved are $g_A = 0$, 0.5, 1.0, 1.5, 2.0, 2.5, and 3.0 g L^{-1} and scattering angles are $\theta = 0$, 20, 40, 60, 80, 100, 120, and 140°.

7-16 Light scattering data obtained from a solution of cellulose acetate in acetone and analyzed with a Zimm plot provide the following results:

$$\text{intercept} = 2.5 \times 10^{-6} \text{ mol g}^{-1}$$

$$\left(\frac{\partial y}{\partial x}\right)_{\theta=0} = 1.32 \times 10^{-5} \text{ mol g}^{-1}$$

$$\left(\frac{\partial y}{\partial x}\right)_{g_A=0} = 2.59 \times 10^{-6} \text{ mol g}^{-1}.$$

The light used in the scattering experiments has the wavelength 436 nm, and the adjustable parameter c used in the plotting has the value 100 cm^3 g^{-1}. Calculate the molar mass, second virial coefficient, and radius of gyration for this macromolecule.

7-17 In this exercise you can demonstrate the effect of the adjustable parameter c in Eq. (7.28) on the appearance and usefulness of a Zimm plot. Make Zimm plots with the program Zimm using the data given in Exercise 7-15 and $c = 10$, 100, 500, 1000, and 10,000 cm^3 g^{-1}. Note changes in the plot caused by changes in c, and decide which one (or ones) of the five plots could be used for an analysis that leads to information on the polymer's molar mass, second virial coefficient, and radius of gyration.

8
Statistical Thermodynamics

The methods of physical chemistry create two worlds—one macroscopic and the other microscopic. The laws of thermodynamics open a door to the macroscopic world, and quantum theory is the key to the microscopic world. This chapter focuses on the bridge between the microscopic and macroscopic worlds erected by the methods of statistical thermodynamics. We begin in Sec. 8.1 with a review of general methods, concluding with statistical equations for calculating internal energy and entropy. A simple example (in the program Mixing) illustrates entropy calculations. The basic tools in the statistical calculation, called partition functions, are introduced in Sec. 8.2 and then calculated (with the programs Zrot, Zvib, and Zelec). In Sec. 8.3 we demonstrate for an ideal gas that all of the thermodynamic state functions—entropies, enthalpies, chemical potentials, and heat capacities—can be calculated from molecular partition functions and spectroscopic data (the program Statcalc). The calculations in Sec. 8.3 are approximate; more refined versions of the same calculations are outlined in Sec. 8.4 (the program Chase). Ideal-gas calculations lead to the methods of statistical chemical thermodynamics described in Sec. 8.5 (the program Statk). The chapter closes with an account of the Pauli principle and the subtleties of nuclear-spin statistics applied to hydrogen and deuterium (programs Cpd2, Cph2, Cphd, S&mud2, S&muh2, S&muhd, and Statg).

8.1 General Methods

Any equilibrated macroscopic system has access to a fantastically large number of quantum states. The energy E_i for one such state is obtained from the energies ε_i of the molecules in the system,

$$E_i = \sum_{\substack{\text{all molecules} \\ \text{in the system}}} \varepsilon_j. \tag{8.1}$$

You will need to remind yourself that in this chapter E_i is an energy of the entire macroscopic system, while the ε_j's are molecular energies. (In previous chapters E_i was used for molecular energies.)

Canonical Ensembles

We call the collection of states that is available to a macroscopic system an ensemble. Particularly important for the thermodynamic description are canonical ensembles for equilibrated, closed, macroscopic systems with the number of molecules N, volume V, and temperature T fixed, but not the energy. Each state included in a canonical ensemble has a probability p_i, which depends just on the energy E_i,

$$p_i = f(E_i) \quad \text{(fixed } N, V \text{ and } T\text{)}, \tag{8.2}$$

with $f(E_i)$ a function of E_i only. We note that, because it is a certainty that a system occupies one of its states, the sum of the p_i's over all of the quantum states equals one,

$$\sum_i p_i = 1. \tag{8.3}$$

The average energy $\bar{E}$ of a closed macroscopic system is simply an ensemble average,

$$\bar{E} = \sum_i p_i E_i,$$

with the summation again covering all of the accessible quantum states. We now build one of the bridges that links the microscopic and macroscopic by identifying the statistically calculated $\bar{E}$ with the thermodynamic quantity U called internal energy,

$$U = \sum_i p_i E_i. \tag{8.4}$$

Corresponding to the energy E_i for the ith quantum state of a macroscopic system, we introduce the entropy S_i, calculated with

$$S_i = -k_B \ln p_i, \tag{8.5}$$

where k_B is Boltzmann's constant, and an average entropy $\bar{S}$ determined by

$$\bar{S} = \sum_i p_i S_i = -k_B \sum_i p_i \ln p_i.$$

We construct another bridge between the macroscopic and microscopic realms by identifying this statistically calculated entropy with the thermodynamic entropy S,

$$S = -k_B \sum_i p_i \ln p_i. \tag{8.6}$$

The two equations (8.4) and (8.6) are all that is needed to begin the development of a statistical thermodynamics for *closed* macroscopic systems.

Microcanonical Ensembles

We model *isolated* macroscopic systems with a different kind of ensemble. This one, called a *microcanonical ensemble* has a fixed number of molecules N, volume V, and energy E (rather than the fixed temperature T in the canonical ensemble). In the microcanonical ensemble, as in the canonical ensemble, probabilities for the quantum states depend only on the energy [Eq. (8.2)]. Thus, with the energy fixed in the microcanonical ensemble all of the states have the same probability, which is written $p_i = 1/\Omega$, with Ω a number later recognized as a measure of disorder. Equations (8.3) to (8.6) apply to the microcanonical ensemble, as well as to the canonical ensemble. Substituting $p_i = 1/\Omega$ in Eq. (8.6), and recalling Eq. (8.3), we find that

$$
\begin{aligned}
S &= -k_B \sum_i p_i \ln(1/\Omega) \\
&= k_B \ln \Omega \sum_i p_i \\
&= k_B \ln \Omega.
\end{aligned}
\tag{8.7}
$$

Equation (8.7) is often interpreted as an expression of the link between entropy and disorder. In that interpretation Ω measures disorder and is calculated as the number of ways a macroscopic system can rearrange itself on a microscopic level without changing its macroscopic state. If, for example, a mixture at equilibrium contains N_A A molecules and N_B B molecules,

$$
\Omega = \frac{N!}{N_A! N_B!},
\tag{8.8}
$$

in which $N = N_A + N_B$. Since factorials are involved and the N's apply to a macroscopic system, Ω is an exceedingly large number. It is usually evaluated with the help of Stirling's approximation,

$$
\ln x! = (x + 0.5) \ln x - x + 0.5 \ln(2\pi),
$$

which reduces to

$$
\ln x! = x \ln x - x
$$

if x is very large.

The program Mixing applies Eq. (8.8) to a mixing process in a system that is not macroscopic in size. The program demonstrates that in such systems the Second Law of Thermodynamics is not precisely obeyed; entropy can decrease in a spontaneous process.

8.2   Partition Functions

System Partition Functions

We return now to canonical ensembles and construct a statistical version of thermodynamics based on the variables N, V, and T. The probability p_i in the canonical ensemble is calculated with

$$p_i = \frac{\exp(-E_i/k_B T)}{Z},$$

(8.9)

in which

$$Z = \sum_i \exp\left(\frac{-E_i}{k_B T}\right),$$

(8.10)

is a *system partition function*. You are reminded again that E_i is the total energy of the macroscopic system in the ith quantum state, and that the summation covers all of the quantum states available to the system.

Molecular Partition Functions

At this point, we could derive equations that express all of the thermo-dynamic state functions in terms of the system partition function Z and its derivatives. It is more convenient, however, to switch from the system parti-tion function Z to a *molecular partition function* which is also a sum of ex-ponential terms, but the energies involved are molecular energies ε_i rather than system energies E_i. One way to define a molecular partition function is

$$z' = \sum_i \exp\left(-\frac{\varepsilon_i}{k_B T}\right),$$

(8.11)

with the summation now covering all of the energy states accessible to an individual molecule. If the summation covers energy levels instead of energy states,

$$z' = \sum_j g_j \exp\left(-\frac{\varepsilon_j}{k_B T}\right),$$

(8.12)

in which g_j is the degeneracy of the jth energy level. If this partition function is multiplied by $\exp(\varepsilon_0/k_B T)$, we obtain a more convenient molecular parti-tion function,

$$z = \exp\left(\frac{\varepsilon_0}{k_B T}\right) \sum_j g_j \exp\left(-\frac{\varepsilon_j}{k_B T}\right)$$

$$= \sum_j g_j \exp\left(-\frac{(\varepsilon_j - \varepsilon_0)}{k_B T}\right).$$

(8.13)

This is the molecular partition function used in the further discussion. As you can see, z is like z' except that all molecular energies are reckoned with respect to the zero-point energy ε_0.

Molecular Partition Functions for Ideal Gases

For an ideal gas the system partition function Z and the molecular partition functions z and z' are related according to

$$Z = \frac{z'^N}{N!} \quad \text{(ideal gas)} \tag{8.14}$$

and

$$Z = \frac{[\exp(-\varepsilon_0/k_\mathrm{B}T)z]^N}{N!} \quad \text{(ideal gas)}, \tag{8.15}$$

where N is the number of molecules in the system.

The energy ε_i of a molecule can always be separated into two added terms, $\varepsilon_j^{\mathrm{tr}}$ for the molecule's translational motion and $\varepsilon_k^{\mathrm{int}}$ for "internal" motion,

$$\varepsilon_i = \varepsilon_j^{\mathrm{tr}} + \varepsilon_k^{\mathrm{int}}. \tag{8.16}$$

At the same time, the molecule's partition function separates into two corresponding multiplied factors,

$$z = z_{\mathrm{tr}}z_{\mathrm{int}}. \tag{8.17}$$

As an approximation, we can go further and separate $\varepsilon_t^{\mathrm{int}}$ for internal motion into added terms for rotational, vibrational, electronic, and nuclear-spin motion,

$$\varepsilon_i^{\mathrm{int}} = \varepsilon_j^{\mathrm{rot}} + \varepsilon_k^{\mathrm{vib}} + \varepsilon_l^{\mathrm{elec}} + \varepsilon_m^{\mathrm{nuc}}. \tag{8.18}$$

Corresponding to this is an approximate factoring of z_{int} into separate rotational, vibrational, electronic, and nuclear-spin partition functions,

$$z_{\mathrm{int}} = z_{\mathrm{rot}}z_{\mathrm{vib}}z_{\mathrm{elec}}z_{\mathrm{nuc}}, \tag{8.19}$$

in which

$$z_{\mathrm{rot}} = \sum_j g_j^{\mathrm{rot}}\left(-\frac{(\varepsilon_j^{\mathrm{rot}} - \varepsilon_0^{\mathrm{rot}})}{k_\mathrm{B}T}\right),$$

and similar statements for z_{vib}, z_{elec}, and z_{nuc}.

Boiling it all down to a sentence, statistical thermodynamics depends on system partition functions Z, system partition functions on molecular partition functions z, molecular partition functions on the factors z_{tr} and z_{int}, representing translational and internal modes of motion, and finally z_{int} on the factors z_{rot}, z_{vib}, z_{elec}, and z_{nuc} representing all the internal modes of motion. The last reduction is an approximation, but we will use it in this and the next section. Sec. 8.4 offers more refined calculations.

Translational Partition Functions

The translational molecular partition function z_{tr} for a system of volume V at temperature T and containing molecules of mass m is

$$z_{tr} = \left(\frac{2\pi m k_B T}{h^2}\right)^{3/2} V. \tag{8.20}$$

This partition function, like others in this chapter, carries no units. But, unlike other partition functions, z_{tr} is an extensive quantity, because of the V factor.

Rotational Partition Functions

Recall that the rotational modes of a molecule are defined with respect to the three principal axes, which we have labeled A, B, and C (Sec. 4.5). Three principal moments of intertia I_A, I_B, and I_C are defined with respect to these axes, and we introduce three parameters $(\theta_{rot})_A$, $(\theta_{rot})_B$, and $(\theta_{rot})_C$, called *characteristic rotational temperatures*,

$$(\theta_{rot})_A = \frac{h^2}{8\pi^2 I_A k_B}$$

$$(\theta_{rot})_B = \frac{h^2}{8\pi^2 I_B k_B} \tag{8.21}$$

$$(\theta_{rot})_C = \frac{h^2}{8\pi^2 I_C k_B}.$$

As the name implies, these parameters have temperature (K) units.

For the *rigid-rotor model* of rotational motion the only contribution to z_{int} made by the ith rotational mode is

$$(z_{rot})_i = \left[\sum_J (2J+1)\exp\left(-\frac{(\theta_{rot})_i J(J+1)}{T}\right)\right]^{1/2}. \tag{8.22}$$

As an approximation the summation in this equation can be evaluated as an integral, with the simple result

$$(z_{rot})_i = \left[\frac{T}{(\theta_{rot})_i}\right]^{1/2}. \tag{8.23}$$

Remember that diatomic and linear polyatomic molecules have only two modes of rotational motion, conventionally labeled $i =$ B and C (Sec. 4.5). For those cases the full molecular rotational partition function z_{rot} is calculated with

$$z_{rot} = \frac{1}{\sigma} \prod_{i=B}^{C} (z_{rot})_i \quad \text{(diatomic and linear polyatomic molecules)}, \quad (8.24)$$

with σ the symmetry number.

Nonlinear polyatomic molecules have three modes of rotational motion, labeled $i =$ A, B and C, and the full molecular rotational partition function is

$$z_{rot} = \frac{\sqrt{\pi}}{\sigma} \prod_{i=A}^{C} (z_{rot})_i \quad \text{(nonlinear, polyatomic molecules)}. \quad (8.25)$$

Characteristic temperatures and symmetry numbers are listed in Table 8.1 for some diatomic molecules and in Table 8.2 for some polyatomic molecules. The program `zrot` calculates rotational partition functions with Eq. (8.24) and both the approximate Eq. (8.23) and the more accurate Eq. (8.22). To run the program enter only the temperature and identify the molecule for which the calculation is to be made. The program automatically supplies data from Tables 8.1 and 8.2 in the file `Chap8.m`.

TABLE 8.1. Characteristic rotational and vibrational temperatures, symmetry numbers and molar masses for some diatomic molecules

Molecule	σ	$M/(\text{g mol}^{-1})$	$\theta_{rot}/\text{K}(*)$	$\theta_{vib}/\text{K}(*)$
$Cl_2(g)$	2	70.906	0.351(2)	805.2(1)
$CO(g)$	1	28.010	2.78(2)	3122(1)
$H_2(g)$	2	2.0159	87.55(2)	6322(1)
$HCl(g)$	1	36.461	15.234(2)	4303(1)
$HF(g)$	1	20.006	30.15(2)	5954(1)
$HI(g)$	1	127.91	9.37(2)	3322(1)
$I_2(g)$	2	253.81	0.0538(2)	308.6(1)
$N_2(g)$	2	28.013	2.8748(2)	3393(1)
$Na_2(g)$	2	45.975	0.223(2)	228.9(1)
$O_2(g)$	2	31.999	2.080(2)	2273(1)
$S_2(g)$	2	64.12	0.425(2)	1044(1)
$NO(g)$	1	30.006	2.452(2)	2739(1)
$OH(g)$	1	17.008	27.15(2)	5373(1)

Sources: Characteristic temperatures calculated from spectroscopic data tabulated by K.P. Huber and G. Herzberg, 1979, and by M.W. Chase et al, 1986.
*Numbers in parentheses denote degeneracies.

TABLE 8.2. Characteristic rotational and vibrational temperatures, symmetry numbers and molar masses for some polyatomic molecules

Molecule	σ	$M/(\mathrm{g\,mol}^{-1})$	$\theta_{\mathrm{rot}}/\mathrm{K}$(*)	$\theta_{\mathrm{vib}}/\mathrm{K}$(*)
H_2O(g)	2	18.015	40.10(1)	5261(1)
			20.88(1)	2294(1)
			13.36(1)	5403(1)
NH_3(g)	3	17.031	13.59(2)	5044(1)
			8.91(1)	1470(1)
				5146(2)
				2433(2)
CH_4(g)	12	16.043	7.54(3)	4196(1)
				2206(2)
				4344(3)
				1879(3)
CO_2(g)	2	44.010	0.561(2)	1997(1)
				960(2)
				3380(1)
C_2H_2(g)	2	26.038	1.693(2)	4852(1)
				2839(1)
				4740(1)
				880(2)
				1049(2)

Sources: Characteristic temperatures calculated from spectroscopic data given by G. Herzberg in *Molecular Spectra and Structure. III*, and by M.W. Chase et al., 1986.
* Numbers in parentheses denote degeneracies.

Vibrational Partition Functions

For this calculation we introduce *characteristic vibrational temperatures*. For the ith vibrational mode the characteristic temperature $(\theta_{\mathrm{vib}})_i$ is calculated with

$$(\theta_{\mathrm{vib}})_i = \frac{hc\tilde{\omega}_i}{k_{\mathrm{B}}}, \qquad (8.26)$$

in which $\tilde{\omega}_i$ is the wavenumber for the ith vibrational mode.

For the *harmonic-oscillator model* the contribution made to z_{vib} by the ith vibrational mode is

$$(z_{\mathrm{vib}})_i = \frac{1}{1 - \exp\left(-\dfrac{(\theta_{\mathrm{vib}})_i}{T}\right)}, \qquad (8.27)$$

and the full vibrational partition function for a molecule with b vibrational modes is

$$z_{\mathrm{vib}} = \prod_{i=1}^{b} (z_{\mathrm{vib}})_i. \qquad (8.28)$$

Characteristic vibrational temperatures for some diatomic and polyatomic molecules are listed in Tables 8.1 and 8.2. The program Zvib calculates vibrational partition functions with Eqs. (8.27) and (8.28).

Electronic Partition Functions

If a molecule has $n + 1$ electronic levels located by the wavenumbers $\tilde{T}_0 \ (= 0), \tilde{T}_1, \ldots, \tilde{T}_n$, the electronic partition function z_{elec} is calculated directly as a sum

$$z_{elec} = \sum_{i=0}^{n} g_i \exp\left(-\frac{hc\tilde{T}_i}{k_B T}\right),$$

$$(8.29)$$

in which $g_0, g_1, \ldots, g_n$ are degeneracies for the electronic levels. The program zelec uses Eq. (8.29) to calculate electronic partition functions.

Nuclear-Spin Partition Functions

Nuclear energy states are separated by very large energies, so the sum in Eq. (8.13) has only a single significant term, that for the ground state,

$$z_{nuc} = g_0,$$

where g_0 is the ground-state nuclear-spin degeneracy for the molecule. If the n nuclei in a molecule have the spins $I_1, I_2, \ldots, I_n$, $g_0 = (2I_1 + 1)(2I_2 + 1) \cdots (2I_n + 1)$, and

$$z_{nuc} = \prod_{i=1}^{n} (2I_i + 1).$$

$$(8.30)$$

Physical chemists customarily ignore this partition function in statistical calculations because its effects cancel in calculations of equilibrium constants and thermodynamic state functions for chemical reactions. There is, however, a subtle connection between rotational and nuclear-spin motion dictated by the Pauli Principle, which cannot be ignored in special cases. That matter is discussed in Sec. 8.6.

8.3   Partition-Function Thermodynamics

As has been mentioned, all of the thermodynamic state functions can be calculated from the system partition function Z and its derivatives. The same statement can be made for the molecular partition function z. The relevant equations for ideal-gas molar entropies $S_m(T)$, molar enthalpies $H_m(T)$, molar heat capacities $C_{Pm}(T)$, and chemical potentials $\mu(T)$ are

$$S_m(T) = R \ln z + RT \left(\frac{\partial \ln z}{\partial T} \right)_V - R \ln L + R \qquad (8.31)$$

$$H_m(T) - H_m(0) = RT^2 \left(\frac{\partial \ln z}{\partial T} \right)_V + RT \qquad (8.32)$$

$$C_{Pm}(T) = 2RT \left(\frac{\partial \ln z}{\partial T} \right)_V + RT^2 \left(\frac{\partial^2 \ln z}{\partial T^2} \right)_V + R \qquad (8.33)$$

$$\mu(T) - H_m(0) = -RT \ln z + RT \ln L. \qquad (8.34)$$

Note that molar enthalpies and chemical potentials are calculated with respect to the zero-point enthalpy $H_m(0)$.

Translational contributions to the thermodynamic state functions are obtained by substituting z_{tr} calculated according to Eq. (8.20) in Eqs. (8.31) to (8.34). Results, after collecting all the constants, are

$$[S_m(T)]_{tr} = R \ln \frac{(T/K)^{5/2} [M/(\text{g mol}^{-1})]^{3/2}}{(P/\text{bar})}$$
$$- 9.5758 \text{ J K}^{-1} \text{mol}^{-1} \qquad (8.35)$$

$$[H_m(T)]_{tr} = \frac{5RT}{2} \qquad (8.36)$$

$$[C_{Pm}(T)]_{tr} = \frac{5R}{2} \qquad (8.37)$$

$$[\mu(T) - H_m(0)]_{tr} = -RT \ln \frac{(T/K)^{5/2} [M/(\text{g mol}^{-1})]^{3/2}}{(P/\text{bar})}$$
$$+ (30.362 \text{ J K}^{-1} \text{mol}^{-1}) \, (T). \qquad (8.38)$$

All of the remaining contributions are due to internal (rotational, vibrational, and electronic) modes of motion. For all of these modes the equations are

$$S_m(T) = R \ln z + RT \left(\frac{\partial \ln z}{\partial T} \right)_V \qquad (8.39)$$

$$H_m(T) - H_m(0) = RT^2 \left(\frac{\partial \ln z}{\partial T} \right)_V \qquad (8.40)$$

$$C_{Pm}(T) = 2RT \left(\frac{\partial \ln z}{\partial T} \right)_V + RT^2 \left(\frac{\partial^2 \ln z}{\partial T^2} \right)_V \qquad (8.41)$$

$$\mu(T) - H_m(0) = -RT \ln z, \qquad (8.42)$$

in which z can be z_{rot}, z_{vib}, or z_{elec}.

For example, in a full statistical calculation of the molar entropy for an ideal gas translational entropy $[S_m(T)]_{tr}$ is calculated with Eq. (8.35), and rotational entropy $[S_m(T)]_{rot}$, vibrational entropy $[S_m(T)]_{vib}$, and electronic entropy $[S_m(T)]_{elec}$ with adaptations of Eqs. (8.39) to (8.42). Then, if the total partition function z reduces to multiplied factors, as in Eqs. (8.17) and (8.19), the total entropy is

$$S_m(T) = [S_m(T)]_{tr} + [S_m(T)]_{rot} + [S_m(T)]_{vib} + [S_m(T)]_{elec}. \qquad (8.43)$$

Enthalpy, heat capacity, and chemical potential calculations are handled similarly.

The program Statcalc does all these calculations for gases in standard states (i.e., for the ideal gas with $P = 1$ bar), assuming in the rotational and vibrational calculations that the rigid-rotor and harmonic-oscillator models are valid. The rotational calculation also uses Eq. (8.23), which evaluates a summation approximately as an integral. The following example shows how to run the program.

Example 8-1. Use the program Statcalc to calculate the standard molar entropy, molar enthaply, molar heat capacity, and chemical potential for $H_2O(g)$ at 1000 K.

Answer. Enter the temperature

$$T = 1000;$$

in Statcalc and identify the component

$$gas = H2O;$$

The program assumes $P = 1$ bar for the standard state, reads the necessary spectroscopic data from the file Chap8.m, and calculates

$$S_m^0(1000) = 232.47 \text{ J K}^{-1} \text{ mol}^{-1}$$

$$H_m^0(1000) - H_m^0(0) = 35.829 \text{ kJ mol}^{-1}$$

$$\mu^0(1000) - H_m^0(0) = -196.64 \text{ kJ mol}^{-1}$$

$$C_{Pm}^0(1000) = 41.027 \text{ J K}^{-1} \text{ mol}^{-1}.$$

8.4   Refinements

The rotational and vibrational parts of the calculations just described are limited by several approximations: the rigid-rotor and harmonic-oscillator models are used and the summation in the Eq. (8.22) is evaluated with an

TABLE 8.3. Spectroscopic parameters for some diatomic molecules

Molecule	$\tilde{\omega}_e/\text{cm}^{-1}$	$\tilde{B}_e/\text{cm}^{-1}$	$\tilde{\omega}_e x_e/\text{cm}^{-1}$	$\tilde{\alpha}_e/\text{cm}^{-1}$	g	$\tilde{T}_e/\text{cm}^{-1}$
$Cl_2(g)$	559.751	0.24415	2.6943	0.0015163	1	0
$CO(g)$	2169.52	1.9302	13.453	0.01746	1	0
$H_2(g)$	4401.21	60.853	121.34	3.062	1	0
$HCL(g)$	2889.59	10.5884	52.06	0.3037	1	0
$HF(g)$	4138.73	20.9555	90.05	0.7958	1	0
$HI(g)$	2309.06	6.512	39.73	0.1715	1	0
$I_2(g)$	214.5481	0.037395	0.61626	0.0001243	1	0
$N_2(g)$	2358.57	1.99824	14.324	0.017318	1	0
$Na_2(g)$	159.11	0.15474	0.72142	0.000868	1	0
$O_2(g)$	1580.1932	1.445622	11.9808	1.59×10^{-7}	3	0
$S_2(g)$	724.67	0.2946	2.836	0.00157	3	0
					2	4700
					1	8500
$OH(g)$	3735.21	18.871	82.81	0.714	2	0
					2	139.7
$NO(g)$	1903.6	1.7042	13.97	0.0178	2	0
					2	121.1

Source: K.P. Huber and G. Herzberg, 1979.

integral in Eq. (8.23). The program Chase does more refined statistical calculations for diatomic molecules by making corrections for rotational stretching, vibrational anharmonicity, and rotational–vibrational interactions. For details on the calculation consult comments in the program and the Pitzer and the Chase et al., references cited.

Input to the program, supplied by the file Chap8.m (Table 8.3) consists of values for the rotational and vibrational spectroscopic parameters $\tilde{\omega}_e$, $\tilde{B}_e$, $\tilde{\omega}_e x_e$, $\tilde{\alpha}_e$ and degeneracies g and wave-numbers $\tilde{T}_e$ for low-lying electronic levels. Example 8-2 shows how to use the program Chase.

Example 8-2. Use the program Chase to calculate the standard molar entropy, molar enthalpy, molar heat capacity, and chemical potential of $N_2(g)$ at 2000 K.

Answer. The procedure for running Chase is like that for Statcalc. Enter the temperature,

$$T = 2000;$$

and identify the component,

$$gas = N2;$$

The program calculates the results listed below and compared with the more approximate results obtained using `Statcalc`.

	Calculated with	
	Chase	Statcalc
$S_m^0(2000)/(\text{J K}^{-1}\,\text{mol}^{-1})$	252.070	251.800
$[H_m^0(2000) - H_m^0(0)]/(\text{kJ mol}^{-1})$	64.800	64.534
$[\mu^0(2000) - H_m^0(0)]/(\text{kJ mol}^{-1})$	−439.340	−439.070
$C_{Pm}^0(2000)/(\text{J K}^{-1}\,\text{mol}^{-1})$	35.958	35.678

8.5   Statistical Chemical Thermodynamics

In addition to entropies, enthalpies, chemical potentials, and heat capacities for individual ideal-gas chemical components, equilibrium constants for entire ideal-gas chemical reactions can be calculated statistically. The key to this calculation is a statistical analog of the activity concept. For a component A, define this activity analog as z_A/N_A, with z_A the molecular partition function for A, and N_A the number of molecules of A in the system. A standard value (i.e., at 1 bar pressure) for the translational contribution is calculated with

$$(z_{tr}^0)_A/N_A = (0.025947)[M_A/(\text{g mol}^{-1})]^{3/2}(T/K)^{5/2}. \qquad (8.44)$$

For the full value of z_A^0/N_A, which is now written f_A^0, we have

$$f_A^0 = (0.025947)[M_A/(\text{g mol}^{-1})]^{3/2}(T/K)^{5/2}(z_{int}^0)_A, \qquad (8.45)$$

where $(z_{int}^0)_A$ includes rotational, vibrational, and electronic contributions.

The quantities f_A^0 give us the mathematical wherewithal to formulate a statistical method for calculating equilibrium constants. An equilibrium constant for the generic ideal-gas reaction,

$$a\text{A}(g) + b\text{B}(g) \rightarrow r\text{R}(g) + s\text{S}(g),$$

is calculated with

$$K = \left[\frac{(f_R^0)^r(f_S^0)^s}{(f_A^0)^a(f_B^0)^b}\right]\exp\left[-\frac{\Delta_r H^0(0)}{RT}\right], \qquad (8.46)$$

in which $\Delta_r H^0(0)$ is the reaction's zero-point enthalpy.

The program `Statcalc` calculates equilibrium constants with Eq. (8.46) and the various equations listed earlier for calculating partition functions. The next example introduces the program.

Example 8-3. You are investigating the possibility of preparing HCl(g) by oxidizing $Cl_2(g)$ at a high temperature with steam,

$$2H_2O(g) + 2Cl_2(g) \rightarrow 4HCl(g) + O_2(g).$$

Calculate an equilibrium constant for this reaction at 800 K and interpret the result. The zero-point enthalpy for the reaction is $\Delta_r H(0) = 109.150$ kJ mol^{-1}.

Answer. Enter the temperature

$$T = 800.;$$

and the zero-point enthalpy

$$deltaHO = 109.150;$$

in Statk. Identify the reaction with

$$reactionList = 4\ HCl + O2 - 2\ H2O - 2\ Cl2;$$

Also enter

$$TMin = 300.;$$

$$Tmax = 1000.;$$

for a plot of $\ln K$ vs T^{-1} in the range TMin to Tmax. The program calculates $K = 0.259$ at 800 K. The reaction is endothermic, so K increases with increasing temperature. At $T = 1000$ K, for example, $K = 8.935$. This may be better for your process, since it gives a higher yield of HCl.

8.6   Nuclear-Spin Statistics

Fermions and Bosons

The Pauli principle informs us that systems of particles with half-integer spin (1/2, 3/2, 5/2, etc.) must be represented by antisymmetric wave functions, and systems of particles with integer spin (0, 1, 2, etc.) by symmetric wave functions. The former are called *fermions* and the latter *bosons*. A familiar example is the electron, which has a spin of 1/2 and is therefore a fermion: electronic wave functions in atoms and molecules are always antisymmetric. The principle applies at *any* level, however, from electrons and nuclei to nucleons and quarks.

Ortho and Para H_2 and D_2

We are concerned in this section with the systems of nuclei found in homo-nuclear diatomic molecules, for example, the hydrogen nuclei in H_2 mole-cules. Their spin is 1/2, they are fermions, and the H_2 nuclear system must have an antisymmetric wave function. Three contributions to this wave function—from rotational, vibrational, and nuclear-spin factors—are im-portant. For H_2 and other homonuclear diatomic molecules, the rotational factors are symmetric if the rotational quantum number J is even, and anti-symmetric if J is odd. We consider only the ground vibrational state ($v = 0$); its contributions are always symmetric. Nuclear-spin states contribute both symmetric and antisymmetric factors, three symmetric and one anti-symmetric. The general rule for a homonuclear diatomic molecule whose nuclei have spin I is that the tally of spin factors is $(I + 1)(2I + 1)$ sym-metric and $I(2I + 1)$ antisymmetric, for a total of $(2I + 1)^2$. Molecules with nuclear spin states represented by symmetric factors are labeled *ortho*, and those with antisymmetric factors *para*. Even though they are chemi-cally the same, ortho and para molecules behave as independent species. Ortho-para interconversion is extremely slow unless an efficient catalyst is present.

Because the nuclei in H_2 are fermions, the overall nuclear wave function for the molecule must be antisymmetric. Ortho H_2 whose wave functions have symmetric nuclear-spin factors, must therefore have antisymmetric rotational factors defined by $J = 1, 3, 5 \ldots$. And para H_2, with antisym-metric nuclear-spin factors, must have symmetric rotational factors defined by $J = 0, 2, 4, \ldots$ This remarkable connection between nuclear-spin and rotational states—forced by the Pauli Principle—is summarized below:

| Molecule type | $J =$ | Symmetry* | | Overall | Degeneracy |
		Nuclear-spin factor	Rotational factor		
Para	$0, 2, 4, \ldots$	a	s	a	$(2J + 1)$
Ortho	$1, 3, 5, \ldots$	s	a	a	$3(2J + 1)$

*s = symmetric; a = antisymmetric.

Degeneracies are noted. Rotational states always have the degeneracy $2J + 1$. Nuclear-spin degeneracies are 3 for ortho H_2 molecules and 1 for para H_2 molecules, as we have seen. The total degeneracy is a product of the nuclear-spin and rotational factors, as shown in the table.

Hydrogen nuclei in deuterium, $^2H_2$ or D_2, have spin 1; they are bosons and therefore the overall nuclear wave function for the molecule must be symmetric. The tally for ortho and para molecules in this case switches the role of rotational states:

Molecule type	$J =$	Symmetry*			Degeneracy
		Nuclear-spin factor	Rotational factor	Overall	
Para	$1, 3, 5, \ldots$	a	a	s	$3(2J + 1)$
Ortho	$0, 2, 4, \ldots$	s	s	s	$6(2J + 1)$

*s = symmetric; a = antisymmetric.

The program Statg calculates and plots nuclear-rotational degeneracies for any homonuclear diatomic molecule, given the spin of the nuclei. Run the program and note the alternating participation of ortho and para molecules. For a given value of J only one nuclear-spin state is allowed, either ortho or para, but not both.

Partition Functions

The point just made by the program Statg, that for homonuclear diatomic molecules a given rotational state allows only one nuclear-spin state, tells us that special rules are needed to formulate nuclear-rotational partition functions. Two summations are required, one for para molecules and another for ortho molecules. For H_2 the nuclear-rotational contribution, call it z_{nucrot} is

$$
z_{nucrot} = \sum_{J=0,2,4\ldots} (2J + 1) \exp\left(-\frac{J(J+1)\theta_{rot}}{T}\right)
$$
$$
+ 3 \sum_{J=1,3,5,\ldots} (2J + 1) \exp\left(-\frac{J(J+1)\theta_{rot}}{T}\right). \qquad (8.47)
$$

The first summation covers para molecules and the second ortho. We have previously ignored nuclear-spin contributions in partition function calculations. That is not possible here; we are calculating a composite rotational and nuclear-spin partition function.

For D_2 the nuclear rotational partition function is calculated with

$$
z_{nucrot} = 6 \sum_{J=0,2,4\ldots} (2J + 1) \exp\left(-\frac{J(J+1)\theta_{rot}}{T}\right)
$$
$$
+ 3 \sum_{J=1,3,5,\ldots} (2J + 1) \exp\left(-\frac{J(J+1)\theta_{rot}}{T}\right). \qquad (8.48)
$$

Here the first summation covers ortho molecules, and the second para.

The programs Cpd2 and Cph2 show how Eqs. (8.47) and 8.48) are applied in the calculation of standard molar heat capacities for $H_2(g)$ and $D_2(g)$

at low temperatures. These programs make separate calculations for the ortho and para components and also for the equilibrium and "normal" ortho-para mixtures. Run the programs, compare the H_2 and D_2 heat capacity plots with each other, and also with the HD heat capacity plot made by the program Cphd. None of the orth-para complications are included in the HD calculations. Why not?

Three more programs, S&mud2, S&muh2, and S&muhd, calculate chemical potentials and standard entropies for D_2, H_2, and HD. In these calculations the conventional subtraction of nuclear-spin contributions is observed.

8.7 Exercises

Partition Functions

8-1 Calculate z_{rot} for $O_2(g)$ at 300K.

8-2 Calculate z_{vib} for $Na_2(g)$ at 500 K.

8-3 Calculate z_{rot} for $S_2(g)$ at 1000 K.

8-4 Calculate z_{vib} for $S_2(g)$ at 1000 K.

8-5 Calculate z_{elec} for Si(g) at 5000 K.

Partition-Function Thermodynamics

8-6 Calculate the standard molar entropy of CO(g) at 298.15 K.

8-7 Calculate the standard molar entropy of $CO_2(g)$ at 500 K.

8-8 Calculate the standard molar entropy of $C_2H_2(g)$ (acetylene) at 800 K.

8-9 Calculate $\mu^0(T) - H_m^0(0)$ for $H_2O(g)$ at 2000 K.

8-10 Calculate $\mu^0(T) - H_m^0(0)$ for $NH_3(g)$ at 1000 K.

8-11 Calculate $\mu^0(T) - H_m^0(0)$ for $O_2(g)$ at 5000 K.

8-12 Revise the program Statcalc so it calculates and prints separately translational, rotational, vibrational, and electronic standard molar entropies. Apply this program to NO(g) at 100, 300, 500, and 1000 K. Note the relative importance of the four contributions to the entropy at the four temperatures.

Refinements

8-13 Use the program Chase to calculate the standard molar entropy $S_m^0(T)$ of CO(g) at 1000 K. Compare this result with the same calculation done by Statcalc.

8-14 Use the program Chase to calculate the standard molar enthalpy $H_m^0(T) - H_m^0(0)$ for $O_2(g)$ at 1000 K. Compare this result with same calculation done by Statcalc.

8-15 Use the program Chase to calculate the standard chemical capacity $\mu(T)^0 - H_m^0(0)$ for $I_2(g)$ at 1000 K. Compare this result with the same calculation done by Statcalc.

8-16 Use the program Chase to calculate the standard molar entropy $C_{Pm}^0(T)$ for OH(g) at 1000 K.

Statistical Chemical Thermodynamics

8-17 Use the program Statk to calculate an equilibrium constant for the reaction

$$2CO(g) + O_2(g) \rightarrow 2CO_2(g)$$

at 900 K. The zero-point enthalpy for the reaction is $\Delta_r H^0(0) = -558.69 \text{ kJ mol}^{-1}$.

8-18 Use the program Statk to calculate an equilibrium constant for the reaction

$$1/2 H_2(g) + 1/2 I_2(g) \rightarrow HI(g)$$

at 700 K. The observed value is $K = 7.42$ and the zero-point enthalpy for the reaction is $\Delta_r H^0(0) = -4.217 \text{ kJ mol}^{-1}$.

8-19 Use the program Statk to calculate an equilibrium constant for the reaction

$$H_2(g) + CO_2(g) \rightarrow CO(g) + H_2O(g)$$

at 900 K. The observed value is $K = 0.46$ and the zero-point enthalpy for the reaction is $\Delta_r H^0(0) = 40.425 \text{ kJ mol}^{-1}$.

8-20 Use the program Statk to calculate an equilibrium constant for the reaction

$$1/2 N_2(g) + 3/2 H_2(g) \rightarrow 2NH_3(g)$$

at 500 K. The zero-point enthalpy for the reaction is $\Delta_r H^0(0) = -38.907 \text{ kJ mol}^{-1}$.

8-21 Use the program Statk to calculate an equilibrium constant for the reaction

$$H_2(g) + 1/2 O_2(g) \rightarrow H_2O(g)$$

at 4500 K. The zero-point enthalpy for the reaction is $\Delta_r H^0(0) = -238.921 \text{ kJ mol}^{-1}$.

8-22 Use a revised version of the program Statk to find the temperature at which the reaction

$$2H_2O(g) + 2Cl_2(g) \rightarrow 4HCl(g) + O_2(g)$$

has the equilibrium constant $K = 1$. At this temperature the reaction "turns around." The zero-point enthalpy for the reaction is $\Delta_r H^0(0) = 109.150 \text{ kJ mol}^{-1}$.

8-23   Use a revised version of the program Statk to find the temperature at which the reaction

$$2CO(g) + O_2(g) \rightarrow 2CO_2(g)$$

has the equilibrium constant $K = 1$. The zero-point enthalpy for the reaction is $\Delta_r H^0(0) = -558.69 \text{ kJ mol}^{-1}$.

Nuclear-Spin Statistics

8-24   Run the program Statg for a homonuclear diatomic molecule whose nuclei have spin 0 (e.g., $^{16}O_2$). Note the connection between rotational states (represented by J) and ortho and para nuclear-spin states.

8-25   Run the program Statg for a homonuclear diatomic molecule whose nuclei have spin 1 (e.g., $^{14}N_2$). Note the connection between rotational states (represented by J) and ortho and para nuclear-spin states.

8-26   Run the program Statg for a homonuclear diatomic molecule whose nuclei have spin 3/2 (e.g., $^{35}Cl_2$). Note the connection between rotational states (represented by J) and ortho and para nuclear-spin states.

8-27   Run the program Cph2. Which of the heat capacity curves plotted by the program would be observed with no catalyst present? Which would be observed with an efficient ortho-para-converting catalyst present?

8-28   Write an adaptation of the program Cpd2 which calculates and prints (does not plot) values of $[C_{Vm}(T)]_{\text{rot}}$ at a given temperature T for ortho and para nuclear-spin states of any homonuclear diatomic molecule whose nuclei are bosons. You will need to increase n, which sets the number of terms in the summation. Have the program make calculations at 1, 10, 100, 200, and 300 K for $^{14}N_2$, whose nuclei have spin 1. Also use the program to make the same calculation for D_2, whose nuclei also have spin 1. Why do N_2 and D_2 behave so differently at low temperatures?

8-29   Write an adaptation of the program Cpd2 which calculates and prints (does not plot) values of $[C_{Vm}(T)]_{\text{rot}}$ at a given tepmerature T for ortho and para nuclear-spin states of any homonuclear diatomic molecule whose nuclei are fermions. You will need to increase n, which sets the number of terms in the summation. Have the program make calculations at 1, 10, 100, 200, and 300 K for $^{35}Cl_2$, whose nuclei have spin 3/2. Also use the program to make the same calculation for H_2. Why do Cl_2 and H_2 behave so differently at low temperatures?

9

Physical Kinetics

Our topic for the remaining three chapters, is kinetics. The term is a broad one with an elusive meaning. For the present the subject is divided just two ways, into physical kinetics and chemical kinetics. Physical kinetics means the study of molecular motion and its strictly physical consequences; that is the topic considered in this chapter. Chemical kinetics is the broad study of rates of chemical processes of all kinds. Its applied and theoretical aspects are discussed in the next two chapters.

Sec. 9.1 takes advantage of a statistical analysis introduced by Maxwell for calculating the distribution of molecular speeds as a function of molar mass and temperature (the programs `Maxwell1` and `Maxwell2`). A detailed picture of two molecules colliding is developed in Sec. 9.2, and the calculation provides data for plots of trajectories of the colliding molecules (the program `Collide`). Sec. 9.3 quotes two of the equations that describe the processes of diffusion, and two programs (`Diffuse1` and `Diffuse2`) demonstrate the progress of diffusion from plane and step sources. The topic in Sec. 9.4 is ions in motion and the calculation of limiting molar conductivities. Two programs (`Lambda` and `Onsager`) provide applications.

9.1 Maxwell's Distribution Function

In one of the earliest efforts to express molecular behavior by statistical methods, Maxwell presented an analysis equivalent to the calculation of the probability $h(v)dv$ that molecules in a system at thermal equilibrium have speeds in the range v to $v + dv$. The distribution function $h(v)$ is defined

$$h(v) = 4\pi \left(\frac{m}{2\pi k_B T}\right)^{3/2} v^2 \exp\left(-\frac{mv^2}{2k_B T}\right), \qquad (9.1)$$

in which m is the mass of the molecules, T is the temperature, and k_B is Boltzmann's constant. This is more conveniently expressed with m replaced by the molar mass $M = Lm$ (L = Avogadro's constant), and k_B by the gas constant $R = Lk_B$,

$$h(v) = 4\pi \left(\frac{M}{2\pi RT}\right)^{3/2} v^2 \exp\left(-\frac{Mv^2}{2RT}\right). \qquad (9.2)$$

This is a probability distribution function with units of reciprocal speed, defined so that $h(v)\,dv$ is a probability, and the integral of $h(v)\,dv$ over all possible speeds from 0 to ∞ is equal to 1,

$$\int_0^\infty h(v)\,dv = 1.$$

The program Maxwell1 calculates $h(v)$ with Eq. (9.2) for a sequence of increasing temperatures requested in the program. For each calculation the program plots a frame that can be selected for an animation, as explained in the program's comments. Run the program and note the broadening of the distribution as the temperature increases.

The program Maxwell2 is similar except that it calculates $h(v)$ for a sequence of increasing molar masses M. Run this program and note the narrowing of the distribution as M increases.

9.2   Molecular Collisions

Maxwell's calculation takes a statistical point of view and therefore provides no pictures of individual molecules. We now zoom in on a single inter-molecular collision and calculate trajectories of the two molecules as they approach each other, collide, and part company. We make the calculation using the classical equations of motion (relying on conservation of energy and angular momentum).

Picture two identical spherical molecules of mass m separated by the distance r, approaching each other in opposite directions (Fig. 9.1). Their initial relative speed is g, and the perpendicular distance b between the two initial trajectories is a collision parameter called the *impact parameter*. The molecules are far enough apart initially (as shown in Fig. 9.1) so they do not

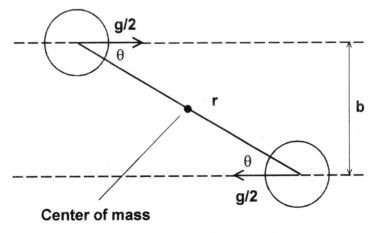

FIGURE 9.1. Two molecules of mass m approaching each other, possibly to meet in a collision.

influence each other. But as they approach, their trajectories change in response to intermolecular forces. If b and g are small enough, a collision occurs.

The kinetic energy $T(r)$ of the system comprising the two molecules is calculated with

$$T(r) = \frac{\mu}{2}\left[\left(\frac{dr}{dt}\right)^2 + r^2\left(\frac{d\theta}{dt}\right)^2\right], \tag{9.3}$$

in which $\mu\ (= m/2)$ is the reduced mass for the two molecules and r and θ are as defined in Figure 9.1. We calculate the potential energy with the *Lennard–Jones function*,

$$V(r) = 4\varepsilon\left[\left(\frac{\sigma}{r}\right)^{12} - \left(\frac{\sigma}{r}\right)^6\right], \tag{9.4}$$

in which ε and σ are empirical parameters. The first term in this function has a repulsion effect and the second an attraction effect.

Initially, with r large enough that $V(r) = 0$, each molecule has the relative energy $\mu g^2/2$. This energy is conserved and at all stages of the intermolecular interaction is equal to the total energy $T(r) + V(r)$,

$$\frac{\mu g^2}{2} = \frac{\mu}{2}\left[\left(\frac{dr}{dt}\right)^2 + r^2\left(\frac{d\theta}{dt}\right)^2\right] + 4\varepsilon\left[\left(\frac{\sigma}{2}\right)^{12} - \left(\frac{\sigma}{r}\right)^6\right]. \tag{9.5}$$

Angular momentum is also conserved during the collision. Initially the approaching molecules have the angular momentum $\mu b g$ and in general the angular momentum is $\mu r^2(d\theta/dt)$, so

$$\mu b g = \mu r^2\frac{d\theta}{dt},$$

or

$$\frac{d\theta}{dt} = \frac{bg}{r^2}. \tag{9.6}$$

Equations (9.5) and (9.6) are the equations of motion needed to solve to determine the trajectory of a collision. The calculation is simplified for programming if we introduce the *reduced quantities*

$$r^* = r/\sigma$$
$$b^* = b/\sigma$$
$$g^* = g\mu^{1/2}/\varepsilon^{1/2}$$
$$t^* = (\varepsilon^{1/2}/\sigma\mu^{1/2})t,$$

so, after some rearrangements, Eqs. (9.5) and (9.6) become

$$\left(\frac{dr^*}{dt^*}\right)^2 = g^{*2}\left(1 - \frac{b^*}{r^{*2}}\right) - 8\left(\frac{1}{r^{*12}} - \frac{1}{r^{*6}}\right) \tag{9.7}$$

$$\frac{d\theta}{dt^*} = \frac{b^* g^*}{r^{*2}}. \tag{9.8}$$

From Eq. (9.7) we extract the function

$$q(r^*) = \left[g^{*2}\left(1 - \frac{b^*}{r^{*2}}\right) - 8\left(\frac{1}{r^{*12}} - \frac{1}{r^{*6}}\right)\right]^{1/2}, \tag{9.9}$$

and note that Eq. (9.7) can be written

$$\frac{dr^*}{dt^*} = \pm q(r^*). \tag{9.10}$$

Both signs are significant in this equation, the minus sign before the molecules reach their distance of closest approach, called the *turning point*, and the plus sign after.

At the turning point r^* has its minimum value r_m^*, calculated with the condition $dr^*/dt^* = 0$, or

$$q(r^*) = 0. \tag{9.11}$$

We rearrange this equation to a polynomial form,

$$g^{*2}r^{*12} - g^{*2}b^{*2}r^{*10} + 8r^{*6} - 8 = 0, \tag{9.12}$$

identify r_m^* as the maximum real root of the polynomial, and calculate the time t_m^* for the molecules to reach the turning point by integrating Eq. (9.10) with the minus sign attached,

$$t_m^* = -\int_{r_0^*}^{r_m^*} \frac{1}{q(r^*)} dr^*, \tag{9.13}$$

where r_0^* is the value of r^* when $t^* = 0$.

The procedure for calculating the full trajectory, followed by the program Collide, is:

1. Calculate r_m^* with Eq. (9.12).
2. Calculate t_m^* with Eqs. (9.9) and (9.13).
3. Solve Eq. (9.10) with the minus sign simultaneously with Eq. (9.8) for $t^* = 0$ to t_m^*, with r_0^* as the initial value of r^*, and

$$\theta_0 = \tan^{-1}\left(\frac{b}{\sqrt{r_0^2 - b^2}}\right)$$

as the initial value of θ.

4. Solve Eq. (9.10) with the plus sign simultaneously with Eq. (9.8) for $t^* = t_m^*$ to $2t_m^*$, with r_m^* as the initial value of r^* and the value of θ calculated at t_m^* in the last step as the initial value of θ.

The program Collide displays the collision trajectory by plotting a vector of length $r/2$ which locates the center of one of the molecules with respect to an origin located at the center of mass for the two molecules. Because the molecules have the same mass m, the center of mass is fixed. Run the program and note the influence of the two parameters b and g on the collision dynamics.

9.3   Diffusion

In a system of uniform composition there is no net transport of molecules in any particular direction. If you construct a plane anywhere in the system and count molecules crossing the plane, there are, on the average, as many molecules crossing the plane in one direction as in the opposite direction. That is the situation assumed in Sec. 9.1. We take a broader viewpoint now and treat the important case of systems with nonuniform composition—not only gaseous but also liquid systems.

If a component is nonuniformly distributed in a system, its concentration is high in some parts of the system and low in others, and its molecules are, on the average, transported in the process of diffusion from regions of high concentration to those of low concentration. Molecules of all sizes are subject to diffusional transport, and the general laws of diffusion are the same in all cases.

One of these laws expresses the expected conclusion that the rate of diffusion in a certain direction depends on the concentration gradient in that direction. We take as a simple example one-dimensional diffusion in the direction x. The diffusion rate is measured as a *molecular flux*: for a component A that flux, call it j_A, is the number of molecules of the component passing through a plane perpendicular to x per unit area per unit time (possible units are cm^{-2} s^{-1}). The molecular flux j_A and concentration gradient dc_A/dx are related by a proportionality,

$$j_A = -D_A \frac{dc_A}{dx}, \tag{9.14}$$

with c_A the molecular concentration and D_A a proportionality factor called a *diffusion coefficient* (units: cm^2 s^{-1} if c_A is expressed in cm^{-3}).

Equation (9.14) is valid if the molecular concentration c_A depends only on the spatial variable x and does not change with time. That is a steady-state condition which obtains if the molecular flux j_A does not change in the x direction. A more general equation includes the time variable,

$$\left(\frac{\partial c_A}{\partial t}\right)_x = D_A \left(\frac{\partial^2 c_A}{\partial x^2}\right)_t. \tag{9.15}$$

This is a one-dimensional version of the *diffusion equation*.

If the molecules are initially located in a thin plane source, and diffuse in the $+x$ and $-x$ directions from the source, Eq. (9.15) has the solution

$$c_A(x, t) = \frac{N}{2\sqrt{\pi D_A t}} \exp\left(-\frac{x^2}{4 D_A t}\right). \tag{9.16}$$

The program Diffuse1 plots this function in an animation. In each frame of the animation $c_A(x, t)$ is plotted as a function of x for a particular value of $D_A t$. The program simulates molecules diffusing from the plane source when the frames are viewed for increasing $D_A t$.

Another useful solution of Eq. (9.15) is

$$c_A(x, t) = \frac{c_{A0}}{2}\left[1 - \mathrm{erf}\left(\frac{x}{4 D_A t}\right)\right], \tag{9.17}$$

which describes diffusion from a step source defined initially by

$$c_A(x, 0) = c_{A0} \quad \text{for} -\infty < x \le 0$$
$$= 0 \quad \text{for } 0 < x < +\infty.$$

The program Diffuse2 calculates frames with Eq. (9.17). When the frames are displayed in an animation they simulate diffusion in the $+x$ direction from the step source.

9.4 Ions in Motion

This section presents some calculations made possible by the theory of electrical conductivity. That theory is simple if the electrolyte whose conductivity is measured is very dilute. In the limiting case of infinite dilution the electrolyte's ions move independently under an applied electrical field, and the limiting molar conductivity Λ_{m0} is equal to the sum of the separate ionic molar conductivities. The situation becomes considerably more complicated in any actual solution involving a finite electrolyte concentration. Then each ion in the solution gathers around itself an ionic atmosphere in which ions of the opposite charge dominate, and an ion must to some extent remain wrapped in this oppositely-charged atmosphere as it moves under the influence of an applied field.

A theory developed by Onsager, Debye, and Hückel calculates the molar conductivity of a strong electrolyte whose molar concentration is C with an approximate equation,

$$\Lambda \cong \Lambda_{m0} - SC^{1/2} \quad \text{(strong electrolye)}, \tag{9.18}$$

in which S is a parameter which also depends on Λ_{m0},

$$S = B_1 \Lambda_{m0} + B_2. \tag{9.19}$$

The further parameters B_1 and B_2 depend on the relative permittivity ε_r, the viscosity coefficient η of the solvent, the temperature T, the stoichiometry of the electrolyte, and the charges carried by the ions.

Consider the case of a symmetric electrolyte, for which the charges z_+ and z_- on the two ions produced by the electrolyte have the same magnitudes: the electrolyte has the formula AX and $|z_+| = |z_-| = z$. Then,

$$B_1 = \frac{qz^3 eF^2}{24\pi\varepsilon_0\varepsilon_r RT}\left(\frac{2}{\varepsilon_0\varepsilon_r RT}\right)^{1/2} \tag{9.20}$$

$$B_2 = \frac{z^3 eF^2}{3\pi\eta}\left(\frac{2}{\varepsilon_0\varepsilon_r RT}\right)^{1/2}, \tag{9.21}$$

where F is Faraday's constant and $q = 2 - \sqrt{2}$.

Equation (9.19) is accurate only for very dilute (but not necessarily infinitely dilute) solutions. The equation can be extended by adding a term that is linear in C,

$$\Lambda_m = \Lambda_{m0} - SC^{1/2} + bC \quad \text{(strong electrolytes)}, \tag{9.22}$$

with b an empirically determined (not a theoretical) parameter. The program Lambda fits molar conductivity data to Eq. (9.22) and calculates the limiting molar conductivity Λ_{m0}, as shown in the next example.

Example 9-1. Data are tabulated below for molar conductivities of aqueous NaCl solutions at 25.00 °C. Use these data and the program Lambda to calculate Λ_{m0} for NaCl at 25.00 °C.

$C/(\text{mol L}^{-1})$	$\Lambda_m/(\text{S cm}^2\,\text{mol}^{-1})$
0.0001	125.56
0.0002	125.21
0.0005	124.50
0.001	123.72
0.002	122.67
0.005	120.58
0.01	118.43
0.02	115.65

Source: *Landoldt–Börnstein*, Vol. II, Part 7, p. 52.

Answer. The program Lambda contains all the data necessary for calculating the constants B_1 and B_2 in Eq. (9.19) for aqueous solutions at 25.00 °C. To run the program enter only the molar conductivity data in the list LmData,

```
LmData = {{.0001, 125.56},
          {.0002, 125.21},
          {.0005, 124.50},
          {.001, 123.72},
          {.002, 122.67},
          {.005, 120.58},
          {.01, 118.43},
          {.02, 115.65}};
```

The program fits these data to Eq. (9.22) and calculates a value for Λ_{m0} (Lmo in the program). The data-fitting task is complicated by the fact that S in Eq. (9.22) cannot be calculated at the outset because it depends on Λ_{m0}, which we need to calculate. That problem is handled in Lambda by *guessing* an initial value for Λ_{m0}, and then, with data fitting, calculating better values of Λ_{m0} and S, followed by another data-fitting step, and so forth. The guessed value of Λ_{m0} can be obtained from the first item in the data list, entered on the first line of code after the data,

```
LmO = 126.;
```

The program calculates $\Lambda_{m0} = 126.452 \ \text{S cm}^2 \ \text{mol}^{-1}$.

For a weak electrolyte (not 100% ionized) Eq. (9.18) must be modified. If the weak electrolyte has the degree of ionization α, αC is substituted for C, the equation becomes

$$\Lambda_{m0} = \alpha(\Lambda_{m0} - S\sqrt{\alpha C}) \quad \text{(weak electrolyte)}, \tag{9.23}$$

and

$$\alpha = \frac{\Lambda_m}{\Lambda_{m0} - S\sqrt{\alpha C}} \quad \text{(weak electrolyte)} \tag{9.24}$$

calculates α.

If a value of the limiting molar conductivity Λ_{m0} is available, Eq. (9.24) can be used to calculate α from molar conductivity data. But the equation has the complication that the unknown quantity α occurs on both sides of the equation. The term containing α on the right is generally small, however, so a first approximation of α can be calculated with $\alpha \cong \Lambda_m/\Lambda_{m0}$. If this value of α is then substituted on the right in Eq. (9.24), a second approximation is obtained. Substituting this on the right in Eq. (9.24) we calculate a third approximation of α, and so forth. This procedure generally converges rapidly to an accurate value of α.

The program `Onsager` implements the procedure outlined and also calculates an equilibrium constant for an ionization of the kind

$$AX \rightleftharpoons A^+ + X^-,$$

according to

$$K = \frac{\gamma_\pm^2 [\alpha(AX)_T]^2}{(AX)_T(1-\alpha)}, \tag{9.25}$$

with $\gamma_\pm$ the mean activity coefficient for the electrolyte. The program calculates $\gamma_\pm$ with the Debye–Hückel limiting law (2.61). `Onsager` is applied to the ionization of acetic acid in the next example.

Example 9-2. Use the program `Onsager` and the acetic acid molar conductivity data (taken at 25.0 °C) quoted below to calculate an ionization constant for acetic acid. The limiting molar conductivity for acetic acid at 25.0 °C is $\Lambda_{m0} = 390.7$ S cm^2 mol^{-1}.

$(CH_3COOH)_T/10^{-4}$	$\Lambda_m/(S\,cm^2\,mol^{-1})$
0.2810	210.350
1.1135	127.750
2.1840	96.493
10.2830	48.146
24.1400	32.217
59.1200	20.962

Answer. Enter the concentration and molar conductivity data,

```
Data = {{.0000281, 210.35},
        {.00011135, 127.75},
        {.0002184, 96.493},
        {.0010283, 48.146},
        {.002414, 32.217},
        {.005912, 20.962}};
```

and the limiting molar conductivity,

```
Lm0 = 390.7;
```

The program calculates the average value $K = 1.752 \times 10^{-5}$ for the acetic acid ionization constant.

9.5 Exercises

Maxwell's Distribution Function (and Others)

9-1 In his 1866 paper, *On the Dynamical Theory of Gases*, Maxwell did not, in fact, mention the distribution function $h(v)$ defined in Eq. (9.2). Instead, he introduced the probability density

$$\frac{1}{\alpha^3 \pi^{3/2}} \exp\left(-\frac{v_x^2 + v_y^2 + v_z^2}{\alpha^2}\right)$$

that a gas molecule has velocity components between

$$v_x \text{ and } v_x + dv_x, \quad v_y \text{ and } v_y + dv_y, \quad v_z \text{ and } v_z + dv_z.$$

Maxwell called the constant α the *modulus of velocity*. In our terms, $\alpha^2 = 2RT/M$, and this function is

$$g(v) = \left(\frac{M}{2\pi RT}\right)^{3/2} \exp\left(-\frac{Mv^2}{2RT}\right). \tag{9.26}$$

Revise the program `Maxwell1` so it calculates and plots this function for values of v ranging from -2000 m s^{-1} to $+2000 \text{ ms}^{-1}$. Suppress the animation and make one plot for $M = 40 \text{ g mol}^{-1}$ and $T = 1000$ K. Compare this plot with that of $h(v)$ and explain the difference.

9-2 Equation (9.2) expresses a *speed* distribution function. It can be converted to a kinetic *energy* distribution function,

$$f(E_k) = 2\pi \left(\frac{1}{\pi RT}\right)^{3/2} E_k^{1/2} \exp\left(-\frac{E_k}{RT}\right), \tag{9.27}$$

where E_k is the molar kinetic energy and $f(E_k)$ is the probability density for kinetic energies between E_k and $E_k + dE_k$. Adapt the program `Maxwell1` so it plots this function at $T = 1000$ K (suppress the animation). Notice that the molar mass M is not needed in this calculation.

9-3 Photons in blackbody radiation are distributed over energy according to a function which has features in common with the Maxwell distribution function $g(v)$. The photon function is usually expressed as a frequency distribution,

$$p(v) = \frac{8\pi h}{c^3} \frac{v^3}{\exp(hv/k_B T) - 1},$$

but it can be converted to a function of the photon molar energy $E = Lhv$,

$$p(E) = \frac{8\pi}{L^4 h^3 c^3} \frac{E^3}{\exp(E/RT) - 1}. \tag{9.28}$$

This function has a maximum value for E_{MostProb} determined by $d\rho(E)/dE = 0$, or

$$\frac{E_{\text{MostProb}}}{RT} = 3\left[1 - \exp\left(-\frac{E_{\text{MostProb}}}{RT}\right)\right],$$

which has the approximate solution

$$E_{\text{MostProb}} = 2.82144RT. \tag{9.29}$$

Write an adaptation of the program `Maxwell1` which plots $\rho(E)$ in an animation covering the temperature range 500 to 2000 K in steps of 100 K.

Molecular Collisions

9-4 If the impact parameter b^* is large enough, two molecules may or may not actually come in contact with each other when they "collide." At low g^* the collision is a contact event, but for higher g^* the molecules pass each other without making contact. Between these two cases there is a special situation in which the collision interaction causes the two molecules to *orbit* each other. Such an orbiting collision is defined by

$$dr^*/dt^* = 0, \tag{9.30}$$

and also by

$$dU^*/dr^* = 0 \quad \text{(maximum)}, \tag{9.31}$$

in which U^* is the *effective potential*,

$$U^* = 4\left(\frac{1}{r^{*12}} - \frac{1}{r^{*6}}\right) + \frac{b^{*2}g^{*2}}{2r^{*2}}, \tag{9.32}$$

including a Lennard-Jones term and also a *centrifugal potential* due to the centrifugal force. This function has a minimum and a maximum, both defined by $dU^*/dr^* = 0$; as Eq. (9.31) indicates, orbiting collisions are defined by the maximum. From Eqs. (9.7), (9.12) and (9.30) we derive

$$g^{*2}r^{*12} - b^{*2}g^{*2}r^{*10} + 8r^{*6} - 8 = 0, \tag{9.33}$$

and, from Eqs. (9.31) and (9.32),

$$g^{*2}b^{*2}r^{*10} - 24r^{*6} + 48 = 0. \tag{9.34}$$

Eliminating $g^{*2}b^{*2}r^{*10}$ between Eqs. (9.33) and (9.34), we have

$$g^{*2}r^{*12} - 16r^{*6} + 40 = 0,$$

which has the solution

$$r_m^* = \left[\frac{16 + \sqrt{256 - 160\,g^{*2}}}{2\,g^{*2}}\right]^{1/6}, \tag{9.35}$$

corresponding to the U^* maximum. Notice that no real value of r_m^* exists, and orbiting collisions are not possible, if $g^* > (256/160)^{1/2} = 1.26491$. To calculate a value of b^* corresponding to g^* we substitute r_m^* in Eq. (9.33) and solve for b^*,

$$b^* = \left[\frac{g^{*2}r_m^{*12} + 8r_m^{*6} - 8}{g^{*2}r_m^{*10}}\right]^{1/2}. \tag{9.36}$$

Write an adaptation of the program Collide that simulates an orbiting collision by including Eqs. (9.35) and (9.36). Input to the program should be a value of g^*. There is no turning point in this case, but instead an "orbiting point," where orbiting begins. Have the program calculate (approximately) t_m, the time elapsed before the molecules reach the orbiting points, then calculate and plot the trajectory before the orbiting point, and finally plot a circle of radius $r_m/2$ simulating the orbits (they are all alike). Run the program for $g^* = 0.1, 0.5, 1.0$ and 1.2. How does the value of g^* affect the orbits? For more on orbiting collisions and other aspects of collision dynamics see Eyring, Lin and Lin.

9-5 Usually the molecules that meet in a collision are deflected, that is, directions of their trajectories before and after the collision are different. Under special conditions, that is not the case. Demonstrate this by running the program Collide for $b^* = 2$ and various values of g^* between 0.3 and 0.4. By trial and error find a value of g^* for which the collision causes no deflection, even though the two molecules contact each other in the collision.

9-6 The program Collide plots the trajectory of one molecule involved in a two-molecule collision. Improve the picture by modifying Collide so it plots trajectories for both of the molecules.

Diffusion

9-7 The program Diffuse1 plots frames for an animation that displays the progress of diffusion for increasing values of the variable $D_A t$. Revise the program so it plots frames for increasing time t rather than $D_A t$. Assume that the diffusion coefficient is $D_A = 10^{-5} \text{ cm}^2 \text{ s}^{-1}$.

9-8 Revise the program Diffuse1 so that, instead of plotting a series of frames, each one of which is a complete plot of $c_A(x, D_A t)$ vs x for a particular value of $D_A t$, the program makes a single plot of $c_A(x, D_A t)$ vs $D_A t$ for a particular x.

9-9 Write a program which makes a three-dimensional plot of $c_A(x, D_A t)$ vs x and $D_A t$ for a plane source.

Ions in Motion

9-10 Data are tabulated below for molar conductivities of aqueous NaI solutions at 25.0 °C. Use these data, the program Lambda and $\lambda_{Na^+} = 50.11 \, S\,cm^2\,mol^{-1}$ for the ionic molar conductivity of Na^+ to calculate the ionic molar conductivity of I^-.

$C/(mol\,L^{-1})$	$\Lambda_m/(S\,cm^2\,mol^{-1})$
0.0005	125.36
0.001	124.25
0.005	121.25
0.01	119.24
0.02	116.70

Source: *Landoldt–Börnstein*, Vol. II, Part 7, p. 53.

9-11 Some electrolytes usually regarded as "strong" are actually to some extent "weak" because of extensive ion-pair formation. Demonstrate this to be the case for $MgSO_4$ in dilute aqueous solutions, using the program Onsager and the molar conductivities tabulated below. The limiting molar conductivity for $MgSO_4$ is $266.1 \, S\,cm^2\,mol^{-1}$.

$(MgSO_4)_T/(10^{-4}\,mol\,L^{-1})$	$\Lambda_m/(S\,cm^2\,mol^{-1})$
0.8098	254.62
1.634	248.54
2.692	242.68
4.297	235.70
6.005	229.84
8.380	223.22

Source: C.W. Davies, 1962, p. 13.

10
Chemical Kinetics in Use

Many chemical and biological processes have measurable rates that can be studied in experiments and modeled mathematically with the methods of chemical kinetics. The strategy of the modeling, sketched in Sec. 10.1, is conceptually simple, and it has been applied far and wide by practitioners of many kinds. Sec. 10.1 applies the kinetic analysis to reacting systems approaching equilibrium and steady-state conditions (the programs Chemkin and Xss). The kineticist's view of more complicated reaction schemes is illustrated in Sec. 10.2, with accounts of the kinetics of catalytic cycles, simplified versions of biochemical cycles, and branching reactions (the programs Catcycle, Cycle, Krebs, and Branch). In Sec. 10.3 the kinetic analysis implied by two mechanisms of polymerization reactions is the topic (the programs Chain and Step). Not only chemists, but biologists and biophysicists (among others, including geologists, chemical engineers, and pharmacologists), have found remarkable uses for the concepts and methods of chemical kinetics. We see an example of biological kinetics in Sec. 10.4 (Biokin) and biophysical kinetics in Sec. 10.5 (Hodgkin).

10.1 Chemical Kinetics

Chemical rate processes are modeled with differential equations, each equation expressing a rate as a function of concentrations of the reactants entering into the process. For example, if the chemical process of interest is a reaction whose stoichiometry is

$$A + B \longrightarrow R, \tag{10.1}$$

the rate of the reaction, expressed as $d[R]/dt$, the rate of formation of the product R, might be calculated with

$$\frac{d[R]}{dt} = k[A][B], \tag{10.2}$$

in which [ ] denotes a molar concentration, and k is a composition-independent, but usually temperature-dependent, "constant" called a *rate constant*.

Chemical statements like (10.1) can be understood on two levels: "$A + B \rightarrow R$" may represent an *elementary reaction* or a *complex reaction*. If (10.1) is elementary it is irreducible; the reactions occurs when molecules of A and B interact directly. But if (10.1) represents the net result of a complex chemical process, it may ignore some of the details such as the occurrence of *intermediates*. For example, A and B may participate in an elementary reaction which produces the intermediate X,

$$A + B \xrightarrow{1} X, \tag{10.3}$$

and then X participates in a second elementary reaction which forms the product R,

$$X \xrightarrow{2} R. \tag{10.4}$$

These two elementary steps constitute the *mechanism* of the reaction. Their net effect is (10.1).

The rate equation (10.2) has an unfortunate ambiguity. It is certainly valid if (10.1) is an elementary reaction, but it may *also* be valid if the reaction has a mechanism expressed by (10.3) and (10.4), and it is even consistent with other mechanisms. If Eq. (10.2), or another equation like it, applies to an elementary reaction we will call it a *rate equation*; if the equation applies to an entire complex reacting system we will call it a *rate law*. In Secs. 10.1–10.3 we will model elementary reactions with rate equations, while in Secs. 10.4–10.5 some complicated biological processes will be modeled with rate laws.

We begin with the simple case of a reversible bimolecular reaction,

$$A + B \underset{2}{\overset{1}{\rightleftarrows}} R + S,$$

with the rate *equations*,

$$\frac{d[A]}{dt} = -k_1[A][B] + k_2[R][S] \tag{10.5}$$

$$\frac{d[B]}{dt} = -k_1[A][B] + k_2[R][S] \tag{10.6}$$

$$\frac{d[R]}{dt} = k_1[A][B] - k_2[R][S] \tag{10.7}$$

$$\frac{d[S]}{dt} = k_1[A][B] - k_2[R][S]. \tag{10.8}$$

This reacting system approaches a chemical equilibrium condition in which the derivatives in Eqs. (10.5) to (10.8) vanish, the concentrations reach their equilibrium values $[A]_e$, $[B]_e$, $[R]_e$, $[S]_e$, and

$$\frac{[R]_e[S]_e}{[A]_e[B]_e} = \frac{k_1}{k_2}, \tag{10.9}$$

where k_1/k_2 is a kinetic evaluation of the equilibrium constant for the reaction.

The program Chemkin integrates Eqs. (10.5) to (10.8) for given values of k_1 and k_2, and also calculates equilibrium concentrations by finding the roots of Eq. (10.9) in combination with the stoichiometric conditions

$$[A]_0 - [A]_e = [B]_0 - [B]_e \tag{10.10}$$

$$[R]_e - [R]_0 = [S]_e - [S]_0 \tag{10.11}$$

$$[A]_e + [B]_e + [R]_e + [S]_e = [A]_0 + [B]_0 + [R]_0 + [S]_0, \tag{10.12}$$

where $[A]_0$, $[B]_0$, $[R]_0$, and $[S]_0$ are initial concentrations.

Many reacting systems are unable to reach a chemical equilibrium condition, but some arrive at a *steady-state condition* instead, which has features in common with equilibrium. We take as an example a reaction scheme in which an intermediate X forms,

$$A + B \underset{2}{\overset{1}{\rightleftharpoons}} X, \tag{10.13}$$

$$X + B \overset{3}{\longrightarrow} R + S. \tag{10.14}$$

The net reaction, not involving the intermediate X, is

$$A + 2B \longrightarrow R + S.$$

Rate equations for this scheme are

$$\frac{d[A]}{dt} = -k_1[A][B] + k_2[X] \tag{10.15}$$

$$\frac{d[B]}{dt} = -k_1[A][B] + k_2[X] - k_3[B][X] \tag{10.16}$$

$$\frac{d[X]}{dt} = k_1[A][B] - k_2[X] - k_3[B][X]. \tag{10.17}$$

If either k_2 or k_3 or both are large compared to k_1, the intermediate X has a very small concentration and sooner or later (usually sooner) reaches a steady-state condition in which $d[X]/dt \cong 0$, and from Eq. (10.17),

$$[X]_{ss} \cong \frac{k_1[A][B]}{k_2 + k_3[B]}, \tag{10.18}$$

where $[X]_{ss}$ denotes a steady-state concentration. In spite of appearances,

$[X]_{ss}$ is not independent of time (as it would be if X were in equilibrium), because [A] and [B] change with time.

This behavior is demonstrated by the program Xss, which calculates and plots $[X]_{ss}$ according to the accurate equations (10.15) to (10.17), and also with the approximate equation (10.18). The program shows that Eq. (10.18) is approximately correct during most of the course of the reaction with $k_1 = 0.1 \, \mathrm{L \, mol^{-1} \, s^{-1}}$, $k_2 = 0.1 \, \mathrm{s^{-1}}$ and $k_3 = 10 \, \mathrm{L \, mol^{-1} \, s^{-1}}$. Run the program for other values of k_1, k_2, and k_3 and verify the same conclusion for any other values of the rate constants satisfying k_2 and/or $k_3 \gg k_1$. If these conditions are not met (e.g., with $k_1 = 0.1 \, \mathrm{L \, mol^{-1} \, s^{-1}}$, $k_2 = 0.1 \, \mathrm{s^{-1}}$ and $k_3 = 0.5 \, \mathrm{L \, mol^{-1} \, s^{-1}}$), Eq. (10.18) is not accurate enough to be useful.

10.2   Complex Reaction Systems

Catalytic Cycles

We look now at some more complicated examples of reacting systems. Consider first a catalytic cycle of reactions,

$$A + X \xrightarrow{1} R + Y \tag{10.19}$$

$$B + Y \xrightarrow{2} S + X. \tag{10.20}$$

The intermediates X and Y are catalysts—they are both produced and consumed in the reactions—and the reaction catalyzed is

$$A + B \longrightarrow R + S.$$

Rate equations for this scheme are

$$\frac{d[A]}{dt} = -k_1[A][X] \tag{10.21}$$

$$\frac{d[B]}{dt} = k_1[B][Y] \tag{10.22}$$

$$\frac{d[X]}{dt} = -k_1[A][X] + k_2[B][Y] \tag{10.23}$$

$$\frac{d[Y]}{dt} = k_1[A][X] - k_2[B][Y]. \tag{10.24}$$

The program Catcycle integrates these equations and plots the results. Run the program, and notice the behavior of the catalysts X and Y; their concentrations change while the reaction proceeds and return to the initial values, as they should for catalysts, when the reaction is completed.

Reaction Cycles

Many biochemical schemes of reactions occur in series, with the products of one reaction serving as the reactants in a subsequent reaction. If the series of reactions also forms a cycle the entire system can maintain a permanent steady-state condition. A simple example (not found in biochemistry) is the scheme

$$A \xrightarrow{1} B$$

$$B \xrightarrow{2} R$$

$$R \xrightarrow{3} S$$

$$S \xrightarrow{3} A,$$

whose rate equations are

$$\frac{d[A]}{dt} = k_4[S] - k_1[A] \qquad (10.25)$$

$$\frac{d[B]}{dt} = k_1[A] - k_2[B] \qquad (10.26)$$

$$\frac{d[R]}{dt} = k_2[B] - k_3[R] \qquad (10.27)$$

$$\frac{d[S]}{dt} = k_3[R] - k_4[S]. \qquad (10.28)$$

Given an initial supply of at least one of the components, this scheme establishes a true steady-state condition in which all of the concentrations are exactly constant, unlike the steady state mentioned in the last section where the concentration of the intermediate X was only approximately constant. After an initial transient period this cyclic scheme does not accomplish a net chemical reaction: it just maintains the components in steady-state concentrations. The program Cycle integrates Eqs. (10.25) to (10.28) and plots concentrations of the four components.

Here is another cyclic reaction scheme, which comes closer to simulating biochemical schemes, although it is still much simpler than the real thing,

$$A + R \xrightarrow{1} B + S$$

$$B \xrightarrow{2} C$$

$$R + C \xrightarrow{3} S + D$$

$$D \xrightarrow{4} A.$$

To keep the system in continuous operation another reaction is included, which produces R at a constant rate from a reactant P whose concentration is large enough to be nearly constant,

$$P \xrightarrow{5} R.$$

This scheme, unlike the last one, does accomplish a net reaction,

$$R \longrightarrow S.$$

Rate equations in this case are

$$\frac{d[A]}{dt} = k_4[D] - k_1[A][R] \tag{10.29}$$

$$\frac{d[B]}{dt} = k_1[A][R] - k_2[B] \tag{10.30}$$

$$\frac{d[C]}{dt} = k_2[B] - k_3[C][R] \tag{10.31}$$

$$\frac{d[D]}{dt} = k_3[C][R] - k_4[D] \tag{10.32}$$

$$\frac{d[R]}{dt} = k_5 - k_1[A][R] - k_3[C][R], \tag{10.33}$$

The program Krebs (so named because the scheme represented is a simplification of the Krebs cycle of biochemical reactions) integrates Eqs. (10.29) to (10.33) and plots concentrations of A, B, C, D and R. Run the program and notice that all of the components eventually reach steady-state concentrations, even the component R, which is being consumed in the cycle. Why is R not depleted?

Branching Reactions

The simple catalytic cycle represented by (10.19) and (10.20) has many elaborations. Consider a scheme involving three catalysts, X, Y and Z, and three reactions,

$$A + X \xrightarrow{1} Y + Z$$

$$B + Z \xrightarrow{2} X + Y$$

$$B + Y \xrightarrow{3} R + X.$$

Two molecules of Y are produced in the first two reactions, and both can react with B in the third. Thus one cycle of the net reaction is

$$A + 3B + X \longrightarrow 2R + 3X,$$

which multiplies the X concentration by three and produces the product R from the reactants A and B. Because of the multiplication this is called a *branching reaction*. If the branching mechanism can be sustained it provides a very efficient reaction path, in fact, too efficient for comfort in some cases when the reaction goes out of control and explodes.

A familiar example is the reaction between hydrogen and oxygen,

$$2H_2(g) + O_2(g) \longrightarrow 2H_2O(g).$$

At high temperatures three free radical components, H, OH, and O, form and participate as X, Y, and Z in the above branching cycle, while O_2, H_2, and H_2O are A, B, and R. Rate equations are

$$\frac{d[H]}{dt} = -k_1[O_2][H] + k_2[H_2][O] + k_3[H_2][OH] \tag{10.34}$$

$$\frac{d[OH]}{dt} = k_1[O_2][H] + k_2[H_2][O] - k_3[H_2][OH] \tag{10.35}$$

$$\frac{d[O]}{dt} = k_1[O_2][H] - k_2[H_2][O] \tag{10.36}$$

$$\frac{d[H_2]}{dt} = -k_2[H_2][O] - k_3[H_2][OH] \tag{10.37}$$

$$\frac{d[O_2]}{dt} = -k_1[O_2][H]. \tag{10.38}$$

The program Branch calculates values for the three rate constants k_1, k_2, and k_3 at a given temperature, then integrates Eqs. (10.34) to (10.38) and plots the results. Run the program and note that both H_2 and O_2 are rapidly depleted, after which the three catalysts, H, OH, and O, have residual steady-state concentrations. That behavior is not actually observed because the free radicals are lost in competing reactions.

10.3 Polymerization Kinetics

In polymerization reactions, small monomer molecules combine to form polymer molecules with large linear or branched chain structures. If a single kind of monomer molecule M is involved the overall polymerization reaction is

$$xM \longrightarrow M_x,$$

with M_x representing a polymer molecule called an *x-mer* whose *degree of polymerization* is x. Monomer molecules enter into the polymerization reaction in two basically different modes called *chain polymerization* and *step polymerization*.

Free-Radical Chain Polymerization

This mechanism, one route for chain polymerization, begins with the formation of initiator radicals $R\cdot$ in the dissociation of initiator molecules I. The rate R_i of the initiation step depends on the concentration [I] of the initiator,

$$R_i = 2fk_d[I], \tag{10.39}$$

where f is an efficiency factor and k_d is a rate constant for initiator dissociation. In the propagation steps of the polymerization, polymer radicals $P\cdot$ form and grow as the monomer molecules M add to initiator radicals and to the polymer radicals themselves. The propagation rate R_p depends on monomer concentration [M], the total polymer radical concentration $[P\cdot]$, and the initiator radical concentration $[R\cdot]$

$$R_p = k_p[M][P\cdot] + k_p[M][R\cdot], \tag{10.40}$$

in which k_p is the propagation rate constant. We assume that chain polymerization terminates when polymer radicals react with each other in combination reactions (neglecting disproportionation reactions), and express the termination rate

$$R_t = 2k_t[P\cdot]^2, \tag{10.41}$$

with k_t a termination rate constant.

Rate equations for the monomer M, polymer radicals $P\cdot$, the initiator I, the initiator radicals $R\cdot$, and the polymer P are

$$\frac{d[M]}{dt} = -R_p = -k_p[M][P\cdot] - k_p[M][R\cdot] \tag{10.42}$$

$$\frac{d[P\cdot]}{dt} = R_i - R_t = 2fk_d[I] - 2k_t[P\cdot]^2 \tag{10.43}$$

$$\frac{d[I]}{dt} = -R_i = -2fk_d[I] \tag{10.44}$$

$$\frac{d[R\cdot]}{dt} = 2fk_d[I] - k_p[M][R\cdot] \tag{10.45}$$

$$\frac{d[P]}{dt} = k_t[P\cdot]^2. \tag{10.46}$$

The average degree of polymerization $\bar{x}$ is calculated with

$$\bar{x} = \frac{-d[M]/dt}{d[P]/dt}$$

$$= \frac{k_p[M][P] + k_p[M][R\cdot]}{k_t[P\cdot]^2}. \tag{10.47}$$

The program Chain integrates Eqs. (10.42) to (10.46) and plots [M], [P·], [I], [R·], [P], and $\bar{x}$ for the first 10 s of a free-radical chain polymerization process. Run the program and note the [M] and [I] are nearly constant for the short time of the calculation, that [P·], [R·] and $\bar{x}$ quickly reach steady-state values, and that [P] continually increases.

Step Polymerization

An example of step polymerization is the formation of a polyester from an hydroxycarboxylic acid. The monomer is HO-R-COOH (R = any divalent group, usually a hydrocarbon) and two monomers react with the elimination of H_2O to form a dimer,

$$HORCOOH + HORCOOH \longrightarrow HO(RCOO)_2H + H_2O.$$

The dimer, also an hydroxycarboxylic acid, can react with another monomer to form a trimer,

$$HORCOOH + HO(RCOO)_2H \longrightarrow HO(RCOO)_3H + H_2O,$$

with another dimer to form a tetramer,

$$H(ORCOO)_2H + HO(RCOO)_2H \longrightarrow HO(RCOO)_4H + H_2O,$$

and so forth.

The essential chemical feature of step polymerization is that any monomer or polymer molecule in the reaction mixture can react with any other monomer or polymer. We assume as an approximation that all of these reactions have the same rate constant, so the kinetics of step polymerization can be treated stochastically.

That is what we do in the program Step. Molecules are chosen randomly for reaction. If M_x and M_y are selected, they form an $(x+y)$-mer M_{x+y}. In the simulation, this reaction is represented by replacing M_x in the list of the system's molecules with M_{x+y} and dropping M_y. The simulation begins with n_0 monomer molecules. When, later, there are n (monomer and polymer) molecules in the system the fractional conversion p of monomer to polymer is defined by

$$p = \frac{n_0 - n}{n_0}. \tag{10.48}$$

Corresponding to this is the average degree of polymerization $\bar{x}$, calculated with

$$\bar{x} = \frac{n_0}{n}. \tag{10.49}$$

The program proceeds to a requested maximum degree of polymerization p_{max} and displays in a bar chart the composition of the reaction mixture. Each molecule is represented by a bar showing its degree of polymerization.

10.4 Biological Kinetics

Many important biochemical components are both supplied and removed in physiological processes. If there are no disturbing influences, the component's rates of input and output to the process are exactly balanced, and the concentration of the component has a steady-state value. Like other steady-state conditions we have met, these are not equilibrium situations. If they were, all related chemical and physical processes would come to a halt and be of no use biologically.

The physiological steady-state concept is exemplified by a kinetic model that simulates the body's handling of glucose. Two components are involved, glucose and insulin, a hormone that facilitates the transport of glucose through cell membranes into the intracellular space where it is utilized. In the kinetic model, rate-law statements are made, values of rate constants are supplied, and a steady-state condition is expressed as a balance between the rates of production and removal of a component.

There are important differences, however, between this example of "biological kinetics" and ordinary chemical kinetics. First, the biological rate laws serve the purposes of simulation only and do not necessarily provide clues concerning the underlying molecular mechanisms. Second, the steady-state conditions reached by the biological components are true balances between input and output rates, not approximate statements contingent on certain components having very small concentrations.

Inputs and Outputs

In the glucose-insulin model, there is one normal glucose input to the extracellular space, from the liver, and three outputs, renal (excretion through the kidney), normal first-order utilization, and the second-order insulin-controlled process that increases the rate of transport of glucose through cell membranes. The renal output is zero unless the glucose exceeds a certain threshold level. The model also shows the effect of artificially raising the glucose concentration temporarily with a glucose infusion. All of these processes are represented in Figure 10.1 with

$$g_1 = \text{rate constant for renal output}$$

$$g_2 = \text{rate constant for first-order utilization}$$

$$g_3 = \text{rate constant for the glucose-insulin interaction}$$

$$R_t = \text{threshold for renal glucose output}$$

$$G_{\text{inf}} = \text{constant rate of glucose infusion}$$

$$G_{\text{liv}} = \text{constant rate of glucose input from the liver.}$$

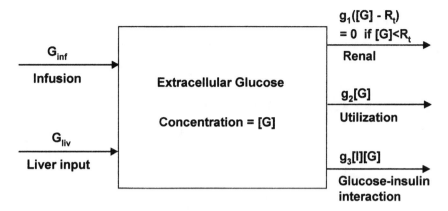

FIGURE 10.1. Glucose inputs to and outputs from the extracellular space.

The model also represents insulin input from the pancreas and output by first-order utilization. The pancreatic input is zero unless the glucose concentration exceeds a threshold level. These processes are represented in Figure 10.2 with

$$i_1 = \text{rate constant for pancreatic input}$$

$$i_2 = \text{rate constant for first-order utilization}$$

$$P_t = \text{threshold for pancreatic input.}$$

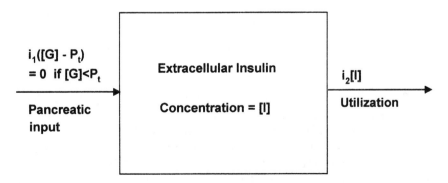

FIGURE 10.2. Insulin inputs to and outputs from the extracellular space.

Rate Laws

This model is expressed mathematically with two rate laws, one for glucose,

$$\frac{d[G]}{dt} = \frac{G_{inf}}{V} + \frac{G_{liv}}{V} - G_{ren} - g_2[G] - g_3[I][G], \tag{10.50}$$

in which V is the total extracellular volume and

$$G_{ren} = g_1([G] - R_t) \quad \text{if } [G] > R_t \tag{10.51}$$
$$= 0 \quad \text{if } [G] \le R_t, \tag{10.52}$$

and another for insulin,

$$\frac{d[I]}{dt} = I_{pan} - i_2[I], \tag{10.53}$$

where

$$I_{pan} = i_1([G] - P_t) \quad \text{if } [G] > P_t \tag{10.54}$$
$$= 0 \quad \text{if } [G] \le P_t. \tag{10.55}$$

Values for the parameters as determined by Stolwijk and Hardy (reported in Mountcastle's *Medical Physiology*) are (with some changes in units):

$$V = 15\,\text{L}$$
$$G_{liv} = 8400 \text{ mg h}^{-1}$$
$$G_{inf} = 80,000 \text{ mg h}^{-1}$$
$$g_1 = 0.480 \text{ h}^{-1}$$
$$g_2 = 0.165 \text{ h}^{-1}$$
$$g_3 = 0.221 \text{ L mg}^{-1} \text{ h}^{-1}$$
$$i_1 = 0.00401 \text{ h}^{-1}$$
$$i_2 = 0.507 \text{ h}^{-1}$$
$$R_t = 2500 \text{ mg L}^{-1}$$
$$P_t = 510 \text{ mg L}^{-1}.$$

The program Biokin integrates Eqs. (10.50) and (10.53) with these data, values for the initial glucose and insulin concentrations, and times t_1 and t_2 between which the glucose infusion is supplied. Run the program and note that [G] and [I] have steady-state values before the glucose infusion and then return to the same steady-state values after the disturbance caused by the infusion. Try other values for the initial concentrations and the rate of infusion G_{inf}, and demonstrate that the system always returns to the same steady-state condition.

10.5   Biophysical Kinetics

Here is one more application of the methods of the chemical kinetics to a biological process. This one concerns Na^+ and K^+ currents across the axon membrane of a nerve cell, resulting in the production of an action potential. We will state the Hodgkin–Huxley equations for these processes and introduce a program Hodgkin which simulates action potentials by solving the equations and plotting the results in an animation.

The two currents mentioned, the sodium current i_{Na^+} and the potassium current i_{K^+}, are calculated with

$$i_{Na^+} = g_{Na^+}(\phi_{Na^+} - \phi_m) \qquad (10.56)$$

and

$$i_{K^+} = g_{K^+}(\phi_{K^+} - \phi_m), \qquad (10.57)$$

where ϕ_m is the potential across the axon membrane, ϕ_{Na^+} and ϕ_{K^+} are equilibrium potentials for Na^+ and K^+, and g_{Na^+} and g_{K^+} are conductances. These look like Ohm's law statements, but they are actually more complicated than that because the conductances g_{Na^+} and g_{K^+} depend on the membrane potential ϕ_m.

The ions flow through channels in the axon membrane, one for Na^+ and an independent one for K^+. The channels behave as if they had gates that open and close under the influence of changes in the membrane potential. When all the gates are open in a channel ions can pass through. When at least one gate is closed, the channel is blocked to the passage of ions.

One way to model axon behavior is to assume that the Na^+ and K^+ channels both have four gates. In the K^+ channels these gates are all of the same kind, which we designate N, and we introduce the variable n for the probability that one N gate is open. Probabilities that two, three, and four of the N gates are open are n^2, n^3, and n^4. Since four gates have to be open to allow passage of K^+ ions, we assume that $g_{k^+} \propto n^4$, or

$$g_{K^+} = \bar{g}_{K^+} n^4, \qquad (10.58)$$

with $\bar{g}_{K^+}$ a proportionality constant.

The Na^+ channels are more complicated. They have two kinds of gates, three of one kind M, each with the probability m for being open, and one of another kind H with the probability h for being open. The probability that all four gates are open, so the channel can pass Na^+ ions, is $m^3 h$, so we assume that

$$g_{Na^+} = \bar{g}_{Na^+} m^3 h, \qquad (10.59)$$

in which $\bar{g}_{Na^+}$ is another constant.

Combining Eqs. (10.56) to (10.59), we have two of the components of the axon membrane current i_m. Two more contributions are important. First, there is a capacitative current i_C, which depends on the membrane's capacitance C_m and also on the rate of change in the membrane potential ϕ_m,

$$i_C = -C_m \frac{d\phi_m}{dt}. \tag{10.60}$$

Second, is a "leakage" current i_L due mostly to passage of Cl$^-$ through the membrane. This current has its own equilibrium potential ϕ_L and conductance g_L (independent of the membrane potential),

$$i_L = g_L(\phi_L - \phi_m). \tag{10.61}$$

Combining all of the currents i_{Na^+}, i_{K^+}, i_L, and i_C, we have the total membrane current i_m,

$$i_m = -C_m \frac{d\phi_m}{dt} + \bar{g}_{Na^+}m^3h(\phi_{Na^+} - \phi_m) + \bar{g}_{K^+}n^4(\phi_{K^+} - \phi_m) + \bar{g}_L(\phi_L - \phi_m). \tag{10.62}$$

The further mathematical burden is lightened by doing the calculations under a *space-clamp condition* in which the membrane current i_m is forced to have a zero value, so Eq. (10.62) becomes

$$\frac{d\phi_m}{dt} = \frac{\bar{g}_{Na^+}m^3h}{C_m}(\phi_{Na^+} - \phi_m) + \frac{\bar{g}_{K^+}n^4}{C_m}(\phi_{K^+} - \phi_m) + \frac{g_L}{C_m}(\phi_L - \phi_m). \tag{10.63}$$

As mentioned, the Na$^+$ and K$^+$ gates are influenced by the membrane potential ϕ_m, and that means the variables n, m, and h depend on the membrane potential. Hodgkin and Huxley expressed that dependence in three first-order rate laws,

$$\frac{dn}{dt} = \alpha_n(1 - n) - \beta_n n \tag{10.64}$$

$$\frac{dm}{dt} = \alpha_m(1 - m) - \beta_m m \tag{10.65}$$

$$\frac{dh}{dt} = \alpha_h(1 - h) - \beta_h h, \tag{10.66}$$

where the α's and β's are rate constants with a sensitive dependence on the membrane potential, calculated by Hodgkin and Huxley using empirical equations formulated for their model preparation, the squid giant axon. The equations are (with ϕ_m measured in mV):

$$\alpha_n = \frac{0.01(\phi_m + 55)}{1 - \exp\left(-\dfrac{\phi_m + 55}{10}\right)} \tag{10.67}$$

$$\beta_n = 0.055\exp\left(-\frac{\phi_m}{80}\right) \tag{10.68}$$

$$\alpha_m = \frac{0.1(\phi_m + 40)}{1 - \exp\left(-\dfrac{\phi_m + 40}{10}\right)} \qquad (10.69)$$

$$\beta_m = 0.108\exp\left(-\frac{\phi_m}{18}\right) \qquad (10.70)$$

$$\alpha_h = 0.0027\exp\left(-\frac{\phi_m}{20}\right) \qquad (10.71)$$

$$\beta_h = \frac{1}{1 + \exp\left(-\dfrac{\phi_m + 35}{10}\right)}. \qquad (10.72)$$

The program Hodgkin solves Eqs. (10.63) to (10.72) simultaneously and simulates action potentials by plotting ϕ_m vs t. The program makes a series of these plots, which can be displayed in an animation. Two important features of action potentials are demonstrated, that they are generated only when the initial membrane potential is above a certain threshold value (about $-60\,\text{mV}$) and that action potentials always have the same shape regardless of the conditions that initiate them.

10.6 Exercises

Chemical Kinetics

10-1 Write a modification of the program Chemkin which includes a third reaction

$$R \xrightarrow{3} Q,$$

in addition to the two already included. Assume that the rate constant for the third reaction is $k_3 = 0.1\ \text{s}^{-1}$. How does the third reaction affect the behavior of the reaction scheme?

10-2 Run the program Xss with the following sets of values of the rate constants k_1, k_2, and k_3. Comment on the validity of the steady-state calculation in each case.

	$k_1/(\text{L mol}^{-1}\,\text{s}^{-1})$	k_2/s^{-1}	$k_3/(\text{L mol}^{-1}\,\text{s}^{-1})$
(a)	0.1	0.1	10.0
(b)	0.1	10.0	0.1
(c)	0.1	10.0	10.0
(d)	10.0	0.1	10.0

10-3 Write a modification of the program Chemkin which calculates and plots A, R, and X concentrations for the reaction scheme

$$A \xrightarrow{1} X \xrightarrow{2} R.$$

Assume $k_1 = 0.5 \text{ s}^{-1}$ and $k_2 = 0.2 \text{ s}^{-1}$ for the rate constants, and $[A]_0 = 0.6 \text{ mol L}^{-1}$, $[X]_0 = 0$ and $[R]_0 = 0$ for the initial concentrations.

10-4 Write a modification of the program Chemkin which calculates and plots A, R, and X concentrations for the reaction scheme

$$A \underset{2}{\overset{1}{\rightleftharpoons}} X \underset{4}{\overset{3}{\rightleftharpoons}} R.$$

Assume $k_1 = 0.5 \text{ s}^{-1}$, $k_2 = 0.2 \text{ s}^{-1}$, $k_3 = 0.1 \text{ s}^{-1}$, and $k_4 = 0.05 \text{ s}^{-1}$ for the rate constants and $[A]_0 = 0.6 \text{ mol L}^{-1}$, $[X]_0 = 0$, and $[R]_0 = 0$ for the initial concentrations. How does the behavior of this scheme differ from that considered in the last exercise?

Complex Reaction Schemes

10-5 Write a revision of the program Cycle which shows the effect of disturbing the reacting system by abruptly changing the concentration of component A. Include three steps in the program:

1. Calculate [A], [B], [R], and [S] for $t_0 = 0$ to $t_1 = 50$ s.
2. Increase [A] to 0.05 mol L^{-1}, but leave [B], [R], and [S] unchanged.
3. Calculate [A], [B], [R], and [S] for $t_1 = 50$ s to $t_2 = 150$ s.

Also have the program calculate algebraically the steady-state concentrations $[A]_{ss}$, $[B]_{ss}$, $[R]_{ss}$, and $[S]_{ss}$.

10-6 Write a revision of the program Krebs which shows the effect of disturbing the reacting system by abruptly changing the concentration of component A. Include three steps in the program:

1. Calculate all components for $t_0 = 0$ to $t_1 = 100$ s.
2. Increase [A] to 0.1 mol L^{-1}, but leave concentrations of all other components unchanged.
3. Calculate concentrations of all components for $t_1 = 100$ s to $t_2 = 300$ s.

Also have the program calculate algebraically the steady-state concentrations $[A]_{ss}$, $[B]_{ss}$, $[C]_{ss}$, $[D]_{ss}$, and $[R]_{ss}$.

10-7 Run the program Branch for $T = 500$, 800, and 1000 K. How does increasing the temperature in this range affect the behavior of the system?

10-8 Consider the catalytic cycle of reactions,

$$A + X \xrightarrow{1} Y + R$$
$$B + Y \xrightarrow{2} 2X + S.$$

Write a program which calculates [A], [B], [X], and [Y] for this system with $k_1 = 0.1 \text{ L mol}^{-1}$ and $k_2 = 1.0 \text{ L mol}^{-1}\text{s}^{-1}$ and $[A]_0 = [B]_0 = [X]_0 = [Y]_0 = 0.1 \text{ mol L}^{-1}$. Run the program and note how the behavior of this system differs from that calculated by the program Catcycle.

10-9 Consider the catalytic cycle of reactions,

$$A + X \xrightarrow{1} 2Y + R$$
$$B + Y \xrightarrow{2} X + S.$$

Write a program which calculates [A], [B], [X], and [Y] for this system with $k_1 = 0.1 \text{ L mol}^{-1}$ and $k_2 = 1.0 \text{ L mol}^{-1}\text{s}^{-1}$ and $[A]_0 = [B]_0 = [X]_0 = [Y]_0 = 0.1 \text{ mol L}^{-1}$. Run the program and note how the behavior of this system differs from that calculated by the program Catcycle.

Polymerization Kinetics

10-10 Run the program Chain with methyl methacrylate as the monomer, for which $k_p = 260 \text{ L mol}^{-1}\text{s}^{-1}$ and $k_t = 1.05 \times 10^7 \text{ L mol}^{-1}\text{s}^{-1}$ at 25 °C.

10-11 Run the program Chain with styrene as the monomer, for which $k_p = 44 \text{ L mol}^{-1}\text{s}^{-1}$ and $k_t = 2.37 \times 10^7 \text{ L mol}^{-1}\text{s}^{-1}$ at 25 °C.

10-12 In our treatment of free-radical chain polymerization reactions we did not include *chain-transfer reactions* in the mechanism. These are termination reactions such as

$$M_m \cdot + TA \longrightarrow AM_m + T \cdot,$$

in which $M_m \cdot$ is a chain radical and TA is *chain transfer agent*. The rate of this reaction is calculated as $k_{tr}[P\cdot][TA]$ and Eq. (10.41) becomes

$$R_t = 2k_t[P\cdot]^2 + k_{tr}[P\cdot][TA], \tag{10.73}$$

so Eqs. (10.46) and (10.47) are

$$\frac{d[P]}{dt} = k_t[P\cdot]^2 + k_{tr}[P\cdot][TA] \tag{10.74}$$

$$\bar{x} = \frac{k_p[M][P\cdot] + k_p[M][R\cdot]}{k_t[P\cdot]^2 + k_{tr}[P\cdot][TA]}. \tag{10.75}$$

Rewrite the program Chain so it includes chain-transfer terms. Run the program with the parameters already included, and in addition $k_{tr} = 100 \text{ L mol}^{-1}$ and $[TA] = 0.01 \text{ mol L}^{-1}$. Note that chain-transfer agents always decrease the average degree of polymerization $\bar{x}$.

10-13 In the bar chart displayed by the program Step each x-mer is listed separately. Adapt Step so it gives a clearer indication of the final distribution of polymer sizes by calculating the final mole fraction f_x for each x-mer with

$$f_x = \frac{n_x}{n}, \tag{10.76}$$

where n_x and n are final values for the number of x-mers and the total number of molecules. Step does its calculation stochastically. The same calculation can be done deterministically using

$$f_x = (1 - p_{max})p_{max}^{x-1}, \tag{10.77}$$

with p_{max} the final conversion of monomer. Write an adaptation of Step which calculates the mole fraction f_x both stochastically and deterministically and displays the calculations together in a plot. For more on Eq. (10.77), see Rudin, 1982, p. 178.

10-14 The program written in the last exercise displays mole fractions f_x for x-mers in a polymer mixture produced by step polymerization. Write a program which calculates the mass fraction w_x, instead of the mole fraction f_x, for each x-mer in the final polymer mixture. Use program (stochastic) data and

$$w_x = \frac{n_x x M}{n_0 M} = \frac{n_x X}{n_0}, \tag{10.78}$$

where M is the molar mass of the monomer and n_0 is the initial number of monomer molecules. This calculation can also be done deterministically with

$$w_x = x(1 - p_{max})^2 p_{max}^{x-1} \tag{10.79}$$

(see Rudin, 1982, p. 178). Have your program compare the stochastic and deterministic calculations.

Biological Kinetics

10-15 Run the program Biokin with the glucose-insulin interaction turned off ($g_3 = 0$) and all other parameters unchanged. How does this affect steady-state values of [G] and [I]?

10-16 Run the program `Biokin` with pancreatic production of insulin turned off ($i_1 = 0$) and all other parameters unchanged. How does this affect steady-state values of [G] and [I]?

10-17 Write a revised version of `Biokin` which calculates steady-state values of [G]. Eliminate the glucose infusion term and suppress the plotting. Have the program solve Eqs. (10.50) and (10.53) with $d[G]/dt = 0$ and $d[I]/dt = 0$, that is,

$$\frac{G_{\text{inf}}}{V} - G_{\text{ren}} - g_2[G] - g_3[I][G] = 0, \tag{10.80}$$

and

$$I_{\text{pan}} - i_2[I] = 0. \tag{10.81}$$

You will need to revise the functions `Grenal` and `Ipancreatic`, representing G_{ren} and I_{pan} in the program, so they depend on [G] rather than t. Run the revised program for the normal case and then for four abnormal cases with: (a) $R_t = 0$; (b) $P_t = 0$; (c) $g_3 = 0$; (d) $i_1 = 0$; and all other parameters unchanged.

Biophysical Kinetics

10-18 Run the program `Hodgkin` with $\phi_{\text{Na}^+} = 20\,\text{mV}$ and all other parameters unchanged. Assume that the initial membrane potential is $\phi_{m0} = -50\,\text{mV}$ and suppress the animation. How does the changed value of ϕ_{Na^+} affect the action potential?

10-19 Run the program `Hodgkin` with $\phi_{\text{K}^+} = -50\,\text{mV}$ and all other parameters unchanged. Assume that the initial membrane potential is $\phi_{m0} = -50\,\text{mV}$ and suppress the animation. How does the changed value of ϕ_{K^+} affect the action potential?

10-20 The Hodgkin–Huxley model simulates the behavior of the axon membrane when it "fires" and produces just one action potential. The model can be modified so it simulates a membrane firing repeatedly, something like the behavior of certain other kinds of excitable membranes, such as heart muscle. Demonstrate this by running the program `Hodgkin` with $\phi_{m0} = -50\,\text{mV}$, $\phi_{\text{K}^+} = -60\,\text{mV}$, $\bar{g}_{\text{Na}^+} = 500\,\text{mS}\,\text{cm}^{-2}$, $t_{\max} = 50\,\text{ms}$, and all other parameters unchanged. Suppress the animation.

10-21 The Hodgkin–Huxley equations (10.63) to (10.72) calculate values of the variable n, which determines the probability that one of the gates in a potassium channel is open. We can use this variable to calculate probabilities for all of the possible configurations of a potassium channel with its 4 N gates. There is a total of 16 of these configurations which we divide into 5 groups designated N_0, N_1, N_2, N_3, and N_4,

N_0: 1 configuration with all gates closed

N_1: 4 configurations with 1 gate open and 3 closed

N_2: 6 configurations with 2 gates open and 2 closed

N_3: 4 configurations with 3 gates open and 1 closed

N_4: 1 configuration with all gates open.

Each group has its own probability:

$$p(N_0) = (1 - n)^4$$
$$p(N_1) = 4(1 - n)^3 n$$
$$p(N_2) = 6(1 - n)^2 n^2 \qquad (10.82)$$
$$p(N_3) = 4(1 - n) n^3$$
$$p(N_4) = n^4.$$

Notice that these probabilities total one, as is necessary because we have accounted for all of the potassium-channel configurations. The N_4 configuration is the only one that conducts K^+ ions. Write an adaptation of the program Hodgkin that calculates these probabilities from $t_{min} = 0$ to $t_{max} = 8.1$ ms and displays the results in a series of bar charts suitable for animation. Assume that the initial membrane potential is $\phi_{m0} = -50$ mV, and leave all other parameters unchanged.

10-22 If you did Exercise 10-21 try this one also, which requires rewriting the program Hodgkin so it calculates probabilities for the 16 configurations allowed by a sodium channel with its 3 M gates and 1 H gate. Divide the configurations into 8 groups,

$H_0 M_0$: 1 configuration with all gates closed

$H_0 M_1$: 3 configurations with the H gate closed, 1 M gate open and 2 closed

$H_0 M_2$: 3 configurations with the H gate closed, 2 M gates open and 1 closed

$H_0 M_3$: 1 configuration with the H gate closed and 3 M gates open

$H_1 M_0$: 1 configuration with the H gate open and all M gates closed

$H_1 M_1$: 3 configurations with the H gate open, 1 M gate open and 2 closed

$H_1 M_2$: 3 configurations with the H gate closed, 2 M gates open and 1 closed

$H_1 M_3$: 1 configuration with all gates open.

Each group has its own probability:

$$p(H_0 M_0) = (1 - m)^3 (1 - h)$$
$$p(H_0 M_1) = 3m(1 - m)^2 (1 - h)$$
$$p(H_0 M_2) = 3m^2 (1 - m)(1 - h)$$
$$p(H_0 M_3) = m^3 (1 - h)$$
$$p(H_1 M_0) = (1 - m)^3 h$$
$$p(H_1 M_1) = 3m(1 - m)^2 h$$
$$p(H_1 M_2) = 3m^2 (1 - m)h$$
$$p(H_1 M_3) = m^3 h.$$

$$(10.83)$$

These probabilities total one because they cover all of the sodium-channel configurations. The last configuration $(H_1 M_3)$ is the only one that allows passage of Na^+ ions. Write an adaptation of the program Hodgkin that calculates these probabilities from $t_{min} = 0$ to $t_{max} = 8.1$ ms and displays the results in a series of bar charts suitable for animation. Assume that the initial membrane potential is $\phi_{m0} = -50$ mV and leave all the other parameters unchanged.

10-23 The Hodgkin–Huxley model is not very sensitive to assumptions concerning the number of potassium and sodium gates. Demonstrate this by rewriting the program Hodgkin for a model that assumes 4 M, 1 H, and 5 N gates. Make the calculation for the initial membrane potential $\phi_{m0} = -50$ mV, but leave all other parameters unchanged. Also make a calculation for a model based on 2 M, 1 H, and 3 N gates.

11
Chemical Kinetics in Theory

One way to approach the theoretical study of chemical reactions is to bring the reactants together in well-defined crossed molecular beams. Then energy allocations to reactant and product molecules can be calculated with a kinematic analysis, as will be shown in Sec. 11.1 and in the program Newton. Beneath the kinematics lies reaction dynamics and its principle theoretical tool, the potential energy surface. A simple method for constructing potential energy surfaces is described in Sec. 11.2, and the program Leps displays graphical representations. The course of a reaction is traced as a path on the potential energy surface, up the valley of the reactants, through a pass at the head of that valley, and down the valley of the products. The configuration of the reacting system at the saddle point in the pass defines the activated complex, which often behaves as if it were in equilibrium with reactant molecules. If so, a statistical method for estimating rate constants introduced by Eyring is applicable. Rudiments of Eyring's theory are sketched in Sec. 11.3 and implemented for unimolecular and bimolecular reactions in the programs Eyring1 and Eyring2. Some features of the further theory of unimolecular reactions are mentioned in Sec. 11.4 and illustrated with the programs Beyer, Rrkm, and Whitten. We saw many examples of steady-state conditions in Chapter 10. Usually, as the name implies, steady states are stable, but not always. In rare cases "steady" states are unstable, and the result can be chemical oscillations. Some examples are discussed in Sec. 11.5 and illustrated with the programs Brussels, Cubecat, Cubictko, Limcycle, Oregon, and Thermkin. A glimpse of the complexities of electrode processes is given in Sec. 11.6 and the programs Butler, Dme, and Rde illustrate. The rate equations of conventional chemical kinetics are accurate when they are applied to large systems. The last section of this chapter (and the book) examines stochastic rate equations that apply to systems of all sizes—large and small—and a program Stokin simulates the chemical kinetics of small systems.

11.1   Reactions in Beams

When two reactants meet each other in a collision event a reaction may take place, but not necessarily. Two factors are important in determining the chemical effectiveness of a collision, the energy of the collision and the orientations of the colliding molecules. In an ordinary "bulb" experiment, colliding molecules approach each other with random orientations and energies covering a broad range. This is the usual situation in kinetic studies of chemical reactions. But some kineticists prefer to design their experiments with the more definite specification of collision parameters allowed in "beam" experiments.

In a *crossed-beam experiment*, the reactants in a bimolecular reaction are carried in collimated beams as high-speed gas-phase molecules at very low pressures. The apparatus in the experiment selects the speeds of the molecules entering the scattering region where the reaction takes place, and also identifies, and measures speeds of, scattered molecules at various angles Θ.

The crossed-beam experiment is easy to understand if you care only about what enters and what leaves the scattering region. Suppose the reaction is

$$A + BC \longrightarrow AB + C,$$

where A, B, and C are atoms. One of the crossed beams contains A and the other BC. The precollision velocities $\mathbf{v}_A$ and $\mathbf{v}_{BC}$ of these reactant molecules are determined in the experiment, and the postcollision velocity $\mathbf{v}'_{AB}$ and the scattering angle Θ of the product AB are also measured. These measurements are made in the *laboratory coordinate system*. But scattering data are easier to comprehend if they are expressed in the *center-of-mass coordinate system*, which travels with the colliding molecules and has its origin placed at the system's center of mass. Precollision velocities of A and BC in the center-of-mass system are $\mathbf{u}_A$ and $\mathbf{u}_{BC}$, and the postcollision velocity of AB is $\mathbf{u}'_{AB}$. The vector diagrams itn Figures 11.1 and 11.2 display these vectors, and

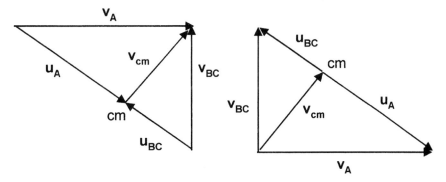

FIGURE 11.1. Vector diagrams representing velocities of the colliding molecules A and BC in the laboratory and center-of-mass coordinate systems. The two diagrams are equivalent physically; the one on the right is the conventional representation.

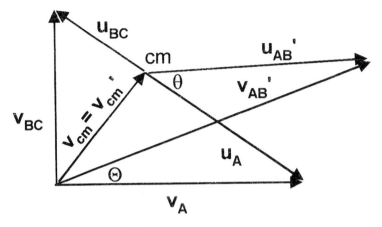

FIGURE 11.2. Newton diagram for the reactive collision $A + BC \rightarrow AB + C$. Note definitions of the scattering angles Θ and θ for the product AB in the laboratory and center-of-mass coordinate systems.

in addition, the velocity $\mathbf{v}_{cm}$ of the center of mass defined by

$$\mathbf{v}_{cm} = \frac{m_A \mathbf{v}_A + m_{BC} \mathbf{v}_{BC}}{M}, \tag{11.1}$$

with $M = m_A + m_{BC}$ the total mass. The vectorial relationship between $\mathbf{u}_A$ and $\mathbf{v}_A$ is

$$\mathbf{v}_A = \mathbf{u}_A + \mathbf{v}_{cm}. \tag{11.2}$$

The precollision relative velocities $\mathbf{v}_A - \mathbf{v}_{BC}$ and $\mathbf{u}_A - \mathbf{u}_{BC}$ are equal; we represent them with $\mathbf{w}$,

$$\mathbf{w} = \mathbf{u}_A - \mathbf{u}_{BC} = \mathbf{v}_A - \mathbf{v}_{BC}, \tag{11.3}$$

and calculate $\mathbf{w}$ with

$$\mathbf{w} = \frac{M \mathbf{u}_A}{m_{BC}}. \tag{11.4}$$

The precollision relative translational kinetic energy E_t is calculated with

$$E_t = \frac{\mu w^2}{2}, \tag{11.5}$$

where

$$\mu = \frac{m_A m_{BC}}{M} \tag{11.6}$$

is the precollision reduced mass.

Now consider the postcollision velocities, represented by $\mathbf{u}'_{AB}$ and $\mathbf{u}'_C$ in the center-of-mass coordinate system, and by $\mathbf{v}'_{AB}$ and $\mathbf{v}'_C$ in the laboratory system. It is demonstrated in Figure 11.2, a *Newton diagram*, that

$$\mathbf{v}'_{AB} = \mathbf{u}'_{AB} + \mathbf{v}'_{cm}, \tag{11.7}$$

where

$$\mathbf{v}'_{cm} = \frac{m_{AB}\mathbf{v}'_{AB} + m_C\mathbf{v}'_C}{M}.$$

Since linear momentum is conserved in the collision, we have

$$m_A\mathbf{v}_A + m_{BC}\mathbf{v}_{BC} = m_{AB}\mathbf{v}'_{AB} + m_C\mathbf{v}'_C,$$

and from this it follows that

$$\mathbf{v}_{cm} = \mathbf{v}'_{cm}.$$

We use $\mathbf{w}'$ to represent either of the relative velocities $\mathbf{v}'_{AB} - \mathbf{v}'_C$ or $\mathbf{u}'_{AB} - \mathbf{u}'_C$,

$$\mathbf{w}' = \mathbf{v}'_{AB} - \mathbf{v}'_C = \mathbf{u}'_{AB} - \mathbf{u}'_C, \tag{11.8}$$

and, in analogy to Eqs. (11.4) and (11.5), arrive at

$$\mathbf{w} = M\frac{\mathbf{u}'_{AB}}{m_C}, \tag{11.9}$$

and

$$E'_t = \frac{\mu' w'^2}{2}, \tag{11.10}$$

where

$$\mu = \frac{m_{AB}m_C}{M} \tag{11.11}$$

is the postcollision reduced mass.

The program Newton implements these calculations. It requires as input data for $\mathbf{v}_A$, $\mathbf{v}_{BC}$, $\mathbf{v}'_{AB}$, Θ (the laboratory scattering angle), m_A, m_B, and m_C; it calculates E_t, E'_t, and θ (the center-of-mass scattering angle); and it draws a Newton diagram.

Example 11-1. The reaction $K + I_2 \rightarrow KI + I$ has been studied by Gillen, Rulis, and Bernstein in crossed-beam experiments. In one series of measurements, the average speed of I_2 molecules in one beam was 172 m s^{-1} and of K in the other was 794 m s^{-1}. The maximum flux of the product was observed at a laboratory scattering angle of $\Theta = 30°$, and the KI speed there was 360 m s^{-1}. Calculate the precollision and postcollision relative translational kinetic energies E_t and E'_t, and the center-of-mass scattering angle θ.

Answer. Enter the given speeds of K, I_2, and KI (in $m\,s^{-1}$) and the scattering angle (in degrees) in Newton,

$$wA = 794.;$$

$$wBC = 172.;$$

$$wAB1 = 360.;$$

$$labTheta = 30.;$$

Also enter molar masses of K, I, and I (in $g\,mol^{-1}$),

$$mA = 39.1;$$

$$mB = 126.9;$$

$$mC = 126.9;$$

The program calculates $E_t = 11.2\ kJ\,mol^{-1}$, $E'_t = 8.3\ kJ\,mol^{-1}$, and $\theta = 20.8°$.

11.2 Potential Energy Surfaces

One of the principal aims of kineticists is to describe the *dynamics* of chemical reactions. In the last section, we made a beginning by describing the *kinematics* of a reaction taking place in the scattering region where two molecular beams cross. We calculated the translational kinetic energy input and output and scattering angle for a reactive collision event. But that description omits an important feature of the dynamics: it says nothing about the *forces* that influence a reactive encounter between molecules. The theory must also include those forces expressed as the potential energies from which they are derived.

As a simple example we again discuss a three-atom reaction of the kind

$$A + BC \longrightarrow AB + C.$$

This reaction begins with the diatomic molecule BC and ends with another diatomic molecule AB. We have seen how to express potential energies of stable diatomic molecules with *Morse functions* (Sec. 4.1). The Morse function used by spectroscopists is given by Eq. (4.3),

$$V(x) = D_e(1 - e^{-\beta x})^2, \tag{11.12}$$

in which x is the displacement of the nuclei in the diatomic molecule from their equilibrium positions (when $V = 0$), β is a constant which can be calculated from spectral data (see the program Morse), and D_e is the asymptotic value of V, approached as $x \to \infty$.

We revise Eq. (11.12) by replacing x with the equivalent $R - R_e$, where R is the internuclear distance and R_e is the equilibrium value of R,

$$V(R - R_e) = D_e[1 - e^{-\beta(R-R_e)}]^2. \tag{11.13}$$

We will also find it expedient to calculate $V(R - R_e) - D_e$ rather than $V(R - R_e)$. Call this quantity $V_M(R - R_e)$,

$$V_M(R - R_e) = V(R - R_e) - D_e = D_e[e^{-2\beta(R-R_e)} - 2e^{-\beta(R-R_e)}], \tag{11.14}$$

and note that $V_M(R - R_e) = -D_e$ when $R = R_e$, and that $V_M(R - R_e) \to 0$ as $R \to \infty$.

With all the parameters in Eq. (11.14) evaluated for BC, and with $R = R_{BC}$, the BC internuclear distance, the equation is a suitable representation for the potential energy of the *reactants* A + BC in the reaction we are considering. [On the reactant side, the internuclear distances R_{AB} and R_{AC} are much larger than R_{BC}, so the potential energy calculated with Eq. (11.14) for the AB and AC interactions can be neglected.] Similarly, Eq. (11.14) with $R = R_{AB}$, and all the parameters evaluated for AB, accounts for the reaction's *products* AB + C.

How do we interpolate between these two extremes to calculate the potential energy of the reacting system when all three atoms are in close proximity? That calculation is never easily done accurately. We use a simple, mostly empirical, calculational method developed in the early days of quantum theory by London, Eyring, and Polyani, and later modified by Sato. The method, labeled LEPS for its authors, was derived in the context of valence bond theory. We quote just the final formulas and show how they are used in the program Leps to make the potential energy calculation for the three-atom reaction.

The central equation in the LEPS three-atom calculation is London's potential energy equation,

$$V_L = Q_{AB} + Q_{BC} + Q_{AC}$$
$$- \{(1/2)[(J_{AB} - J_{BC})^2 + (J_{BC} - J_{AC})^2 + (J_{AC} - J_{AB})^2]\}^{1/2}, \tag{11.15}$$

in which the Q's and J's, called Coulomb and exchange integrals, are valence-bond counterparts of the Coulomb and bond integrals of molecular-orbital theory. Sato modified Eq. (11.15) by including another empirical parameter K in a denominator factor,

$$V_S = \frac{V_L}{1 + K}, \tag{11.16}$$

where V_S is Sato's version of the three-atom potential energy. Equations for the Coulomb and exchange integrals, also involving K, are

$$Q = \frac{D_e}{4}[(3 + K)e^{-2\beta(R-R_e)} - (2 + 6K)e^{-\beta(R-R_e)}] \tag{11.17}$$

$$J = \frac{D_e}{4}[(1 + 3K)e^{-2\beta(R-R_e)} - (6 + 2K)e^{-\beta(R-R_e)}]. \tag{11.18}$$

Equation (11.15) requires calculation of Coulomb and exchange integrals for three possible diatomic combinations, AB, AC, and BC. That means the independent variables in the calculation are the three internuclear distances R_{AB}, R_{AC}, and R_{BC}. One of these variables can be eliminated if we assume a fixed geometry for the transition states. The simplest possibility is linear transition states for which

$$R_{AC} = R_{AB} + R_{BC}. \tag{11.19}$$

11.3 Activated-Complex Theory

In its simplest version activated-complex theory begins with three fundamental assumptions:

- That the reactant molecules, product molecules, and activated complexes are all in thermal equilibrium, that is, they are distributed among their quantum states according to the Boltzmann distribution law.
- That in a reaction forming an activated complex $X^{\ddagger}$,

$$A + B \rightarrow X^{\ddagger} \rightarrow R + S,$$

the reactants A and B, and the activated complex $X^{\ddagger}$, appear to be in chemical equilibrium, so the concentration ratio $[X^{\ddagger}]/[A][B]$ can be assumed to have its equilibrium value. This a vital assumption because it permits the use of partition functions to calculate the concentration ratio (Sec. 8.5).
- That the forward rate of the reaction is equal to the rate of crossings of activated complexes along the reaction path through the saddle point on the reaction's potential energy surface between the reactant and product valleys. This assumption, which has its exceptions, requires that the reaction path traverse the saddle point only once; meandering paths with multiple crossings are not allowed.

These assumptions lead finally to the Eyring equation for calculating rate constants,

$$k = (Lk_B T/h)(q_{\ddagger}/q_A q_B) \exp(-E_0^{\ddagger}/RT) \quad \text{(bimolecular reactions),} \tag{11.20}$$

in which the q's are partition functions expressed per unit volume and $E_0^{\ddagger}$ is the reaction's threshold energy. In Sec. 8.5, we used partition functions defined z_A/N_A for a component A. In Eq. (11.20) they are replaced by partition function of the kind $q_A = z_A/V$, calculated with

$$q_A = (1.8793 \times 10^{23} \text{ L}^{-1})[M_A/(\text{g mol}^{-1})](T/\text{K})^{3/2}(z_{int})_A. \tag{11.21}$$

[Do not confuse L in Eq. (11.20) with L in Eq. (11.21); the former is Avogadro's constant and the latter the liter volume unit.] Thus q_A has the units L^{-1} and $q_{\ddagger}/q_A q_B$ the units L. Because the factor $Lk_B T/h$ has $\text{mol}^{-1} \text{ s}^{-1}$ units, we see that Eq. (11.20) calculates k in the usual units for a bimolecular reaction, $\text{L mol}^{-1} \text{ s}^{-1}$.

Equation (11.2) calculates rate constants for bimolecular reactions. The same analysis applies to unimolecular reactions, except that the partition-function ratio $q_{\ddagger}/q_A q_B$ is replaced by $q_{\ddagger}/q_A$,

$$k = (Lk_B T/h)(q_{\ddagger}/q_A)\exp(-E_0^{\ddagger}/RT) \quad \text{(unimolecular reactions)}. \quad (11.22)$$

The program Eyring1 uses Eqs. (11.21) and (11.22) to calculate rate constants for unimolecular reactions, and Eyring2 does the same thing for bimolecular reactions with Eqs. (11.20) and (11.21).

Example 11-2. Consider the bimolecular gas-phase reactions,

$$H + HBr \longrightarrow H_2 + Br,$$

which you can assume proceeds through the linear activated complex H-H-Br, whose bond distances are $R_{HH} = 0.90\,\text{Å}$ and $R_{HBr} = 1.50\,\text{Å}$. The bond distance in HBr is 1.413 Å, and the single vibrational mode for HBr has the wavenumber 2560 cm^{-1}. Vibrational modes for the activated complex have the wavenumbers 1313, 540 and 540 cm^{-1}. The threshold energy for the reaction is 5.0 kJ mol^{-1}. Estimate the rate constant for the reaction at 300 K.

Answer. The data supplied have been converted to rotational and vibrational characteristic temperatures for HBr and the activated complex, and the results entered in the data file Chap11.m, read by the Eyring programs. To finish the calculation using Eyring2, we need only enter in reactionList the "reaction" that forms the activated complex,

```
ReactionList = HHBrac - H - HBr;
```

Note that the suffix ac denotes an activated complex. Also enter the temperature for which the calculation is to be made,

```
T = 300.;
```

and the value of the reaction's threshold energy (in J mol^{-1}),

```
Edagger0 = 5000.;
```

The program calculates $k = 2.4 \times 10^9$ L mol^{-1} s^{-1}. A measured value is $k = 3 \times 10^9$ L mol^{-1} s^{-1}. Considering the uncertainties, the agreement is good.

11.4   Unimolecular Reactions

Unimolecular reactions are not strictly unimolecular. A reactant molecule acquires energy in a bimolecular interaction with another molecule M, and then the energized A molecule forms products in a unimolecular step. The energized molecule can also lose energy to M molecules. The full mechanism is

$$A + M \xrightarrow{1} A^*(E_i) + M \quad \text{Rate constant} = k_i$$
$$A^*(E_i) + M \xrightarrow{2} A + M \quad \text{Rate constant} = k_2$$
$$A^*(E_i) \xrightarrow{3} \text{Products} \quad \text{Rate constant} = k(E_i),$$

where $A^*(E_i)$ is an energized reactant molecule with energy E_i. The rate constants $k(E_i)$ for the product-forming reaction in Step 3 are called *microscopic rate constants*. This mechanism leads to the rate equation

$$\text{Rate} = \sum_i \frac{k(E_i)[A^*(E_i)]}{1 + k(E_i)/k_2[M]}. \tag{11.23}$$

The number of energy states available to reacting molecules is enormous, large enough so E_i can be treated as a continuous variable E. Then the microscopic rate constants $k(E_i)$ become a continuous function $k(E)$ and the concentration of energized molecules $[A^*(E_i)]$ is expressed $P(E)[A]dE$, where $P(E)dE$ is the fraction of the total concentration of molecules $[A]$ having energies in the range $E + dE$. The summation in Eq. (11.23) becomes an integral,

$$\text{Rate} = \int_{E_0}^{\infty} \frac{k(E)P(E)[A]dE}{1 + k(E)/k_2[M]}, \tag{11.24}$$

in which E_0 is the threshold above which the reaction can take place [i.e., $k(E) = 0$ for $E < E_0$].

Further theoretical progress is made by introducing the continuous density of states $N(E)$, which counts the number of states available per unit of energy in the energy range E to $E + dE$. Not all of these states are occupied by molecules. The Boltzmann distribution function $\exp(-E/RT)/z$ (z is a molecular partition function; see Sec. 8.2) determines the fraction of the states counted by $N(E)$ that is actually occupied. Multiplying $N(E)$ by the Boltzmann factor, we obtain the function $P(E)$ in Eq. (11.24),

$$P(E) = \frac{N(E)\exp\left(-\dfrac{E}{RT}\right)}{z}. \tag{11.25}$$

From Eqs. (11.24) and (11.25), we see that the density of states $N(E)$ and the microscopic rate constants $k(E)$ are the keys to the calculation. The *RRKM theory*, developed by Rice and Ramsperger, Kassel, and most recently, Marcus, calculates the microscopic rate constants $k(E)$ in a statistical manner. The theory assumes that on the time scale of the reaction the reactant's internal molecular energy is rapidly and randomly distributed among all of the molecule's vibrational and rotational states. If the energy E is greater than the threshold energy E_0 for the reaction, this distribution eventually concentrates energy in the right molecular location to form the activated complex, and the reaction can proceed to products.

This statistical picture, and the methods of activated-complex theory, lead to a simple equation for calculating microscopic rate constants,

$$k(E) = G^{\ddagger}(E - E_0)/hN(E), \qquad (11.26)$$

where h is Planck's constant, $N(E)$ is the density of states for the reactant and $G^{\ddagger}(E - E_0)$, a sum of states, covers all the states available to the activated complex in the energy range E_0 to E.

The program Rrkm calculates microscopic rate constants according to Eq. (11.26), Beyer calculates $N(E)$ and $G(E)$ for vibrational states with a direct count algorithm, and Whitten does the same calculation with an analytical procedure.

11.5   Oscillating Reactions

In Sec. 10.2, we discussed chain reactions which are unstable and possibly explosive because the free radical intermediates that propagate the chain cannot be maintained in steady-state concentrations; the radical concentrations increase exponentially and the system can explode.

We now consider some exotic reacting systems that are unstable for more subtle reasons. In these systems steady states for the chain-carrying intermediates exist—at least they can be calculated—but may not be worthy of the name. These "steady" states may be unsteady and unstable. If the system is prepared in such a state it cannot remain there: concentrations of the intermediates diverge from the "steady" values, either monotonically or in oscillations.

The Brusselator

Features of steady-state instabilities and chemical oscillations are most easily appreciated by studying an example. Consider the *Brusselator model*, so called because it was invented by Prigogine and coworkers in Brussels. The reaction scheme for this system is

$$A \xrightarrow{1} X$$
$$B + X \xrightarrow{2} R + Y$$
$$Y + 2X \xrightarrow{3} 3X \qquad (11.27)$$
$$X \xrightarrow{4} S,$$

in which A, B, R, and S are reactants and products in the overall stoichiometry and X and Y are intermediates.

Reaction 3 is like the chain-carrying steps that contribute to explosion kinetics. Here we call them *autocatalytic steps*, and find that their kinetic effect is destabilizing but not catastrophic. Reaction 3 exhibits *cubic auto-catalysis* because the reaction is kinetically third order. To diminish some of the calculational difficulties, we assume that the reactant concentrations [A] and [B] have fixed values.

We write the rate equations for the Brusselator (assuming that [A] and [B] are constant),

$$\frac{d[X]}{d\tau} = \frac{k_1[A] - k_2[B][X] + k_3[X]^2[Y] - k_4[X]}{k_4} \qquad (11.28)$$

$$\frac{d[Y]}{d\tau} = \frac{k_2[B][X] - k_3[X]^2[Y]}{k_4}, \qquad (11.29)$$

in which $\tau = k_4 t$ is a unitless time-related variable. The steady-state conditions for [X] and [Y] are $d[X]/d\tau = 0$ and $d[Y]/d\tau = 0$, and they permit calculation of the steady-state concentrations $[X]_{ss}$ and $[Y]_{ss}$,

$$[X]_{ss} = \frac{k_1[A]}{k_4} \qquad (11.30)$$

$$[Y]_{ss} = \frac{k_2 k_4[B]}{k_1 k_3[A]}. \qquad (11.31)$$

Now consider the stability of these steady states. Our test for stability is an easy one to apply. Imagine that the two concentrations [X] and [Y] are temporarily displaced small amounts $\delta[X]$ and $\delta[Y]$ from the steady-state values $[X]_{ss}$ and $[Y]_{ss}$. If [X] and [Y] respond to this disturbance by returning to $[X]_{ss}$ and $[Y]_{ss}$, the steady state is a stable one. If, on the other hand, the disturbance causes [X] and [Y] to diverge further from $[X]_{ss}$ and $[Y]_{ss}$, the steady state is unstable.

These responses to the displacements $\delta[X]$ and $\delta[Y]$ are calculated by the quantities

$$\left(\frac{\partial[\dot{X}]}{\partial[X]}\right)_{ss} \delta[X] \quad \text{and} \quad \left(\frac{\partial[\dot{Y}]}{\delta[Y]}\right)_{ss} \delta[Y],$$

where [$\dot{X}$] and [$\dot{Y}$] are shorthand notations for $d[X]/d\tau$ and $d[Y]/d\tau$, and the partial derivatives are evaluated for steady-state conditions, as indicated. The rules for stability and instability of a steady state depend on the sign of the sum of the two partial derivatives: if

$$\left(\frac{\partial[\dot{X}]}{\partial[X]}\right)_{ss} + \left(\frac{\partial[\dot{Y}]}{\partial[Y]}\right)_{ss} < 0$$

the steady state is stable, and if

$$\left(\frac{\partial[\dot{X}]}{\partial[X]}\right)_{ss} + \left(\frac{\partial[\dot{Y}]}{\partial[Y]}\right)_{ss} \geq 0$$

it is unstable. For the Brusselator, the stability-instability criterion depends on

$$\left(\frac{\partial[\dot{X}]}{\partial[X]}\right)_{ss} + \left(\frac{\partial[\dot{Y}]}{\partial[Y]}\right)_{ss} = \frac{k_2[B]}{k_4} - \frac{k_1^2 k_3[A]^2}{k_4^3} - 1. \tag{11.32}$$

The program Brussels integrates Eqs. (11.28) and (11.29), plots [X] and [Y] vs τ, and calculates a value for the "stability parameter" $(\partial[\dot{X}]/\partial[X])_{ss} + (\partial[\dot{Y}]/\partial[Y])_{ss}$ according to Eq. (11.32). Run the program for several cases: [A] = [B] = 0.01 mol L^{-1};   [A] = [B] = 0.02 mol L^{-1};   [A] = 0.01 mol L^{-1}, [B] = 0.02 mol L^{-1}; and [A] = 0.02 mol L^{-1}, [B] = 0.01 mol L^{-1}. Note that the stability parameter calculated by Eq. (11.32) correctly predicts stable and unstable steady states.

In cases where [X] and [Y] oscillate the oscillations appear to settle into a pattern of constant amplitude and period. The program Limcycle demonstrates this by plotting [X] vs [Y] for the Brusselator. Regardless of the initial values of [X] and [Y], the plot always leads to the same closed loop called a *limit cycle*. If [A] and [B] do not change, the system traces the limit cycle forever. Run Limcycle several times with different initial values of [X] and [Y].

Another Cubic Autocatalator

Here is another reaction scheme with a cubic autocatalytic step,

$$\begin{aligned} A &\xrightarrow{1} X \\ X &\xrightarrow{2} Y \\ X + 2Y &\xrightarrow{3} 3Y \\ Y &\xrightarrow{4} R. \end{aligned} \tag{11.33}$$

Assuming that [A] has a fixed value, the rate equations are

$$\frac{d[X]}{d\tau} = \frac{k_1[A] - k_2[X] - k_3[X][Y]^2}{k_4} \tag{11.34}$$

$$\frac{d[Y]}{d\tau} = \frac{k_2[X] + k_3[X][Y]^2 - k_4[Y]}{k_4}, \tag{11.35}$$

in which $\tau = k_4 t$. An analysis like that outlined for the Brusselator model leads to

$$\left(\frac{\partial[\dot{X}]}{\partial[X]}\right)_{ss} + \left(\frac{\partial[\dot{Y}]}{\partial[Y]}\right)_{ss} = -c_1 - c_2 + \frac{c_2 - c_1}{c_2 + c_1}, \tag{11.36}$$

where $c_1 = k_2/k_4$ and $c_2 = k_1^2 k_3[A]^2/k_4^3$, for the stability parameter.

The program Cubecat integrates Eqs. (11.34) and (11.35) and plots [X] and [Y] vs τ.

The Oregonator

The examples of oscillating chemical reactions mentioned so far have concerned systems which have no counterpart in actual reacting systems. We turn now to a more realistic system called the *Oregonator model* (developed by Noyes and colleagues at the University of Oregon) which accurately models the much-studied *Belousov–Zhabotinsky (or BZ) reaction.*

The BZ reaction brominates malonic acid (MA) with Br^- and the Ce^{3+}/Ce^{4+} couple as catalysts. The overall stoichiometry is (approximately)

$$2BrO_3^- + 3MA + 2H^+ \xrightarrow{Br^-, Ce^{4+}} 2BrMA + 3CO_2 + 4H_2O.$$

The Oregonator model represents the BZ reaction with the following scheme:

$$
\begin{aligned}
A + Y &\xrightarrow{1} X + R \\
X + Y &\xrightarrow{2} 2R \\
A + X &\xrightarrow{3} 2X + 2Z \\
2X &\xrightarrow{4} A + R \\
B + Z &\xrightarrow{5} 1/2Y,
\end{aligned}
\tag{11.37}
$$

where $A = BrO_3^-$, $B = MA + BrMA$, $R = HOBr$, $X = HBrO_2$, $Y = Br^-$, and $Z = Ce^{4+}$. These reactions are not necessarily elementary; they omit some of the intermediates. Rate laws for this scheme are

$$\frac{d[X]}{d\tau} = \frac{k_1[A][Y] - k_2[X][Y] + k_3[A][X] - 2k_4[X]^2}{k_5[B]} \tag{11.38}$$

$$\frac{d[Y]}{d\tau} = \frac{-k_1[A][Y] - K_2[X][Y] + (1/2)k_5[B][Z]}{k_5[B]} \tag{11.39}$$

$$\frac{d[Z]}{d\tau} = \frac{2k_3[A][X] - k_5[B][Z]}{k_5[B]}, \tag{11.40}$$

where $\tau = k_5[B]$ is another unitless time-related variable.

The program Oregon integrates Eqs. (11.38) to (11.40) and plots $10[X]$, $[Y]$, and $[Z]$ vs τ.

Thermokinetic Oscillators

Chemical engineers are familiar with systems involving exothermic reactions which "catalyze" themselves because they generate thermal energy more rapidly than it can be conducted away from the reactor in which the reaction takes place. The temperature rises and the reaction rate increases because of the exponential (Arrhenius) dependence of the rate constant on temperature. This is a case of *exponential autocatalysis*, and it can lead to unstable steady-state behavior and oscillations, both chemical and thermal. A minimal scheme that can oscillate thermochemically is

$$
\begin{aligned}
A &\xrightarrow{1} X \\
X &\xrightarrow{2} R,
\end{aligned}
\tag{11.41}
$$

in which Reaction 2 has an appreciable reaction enthalpy and activation energy.

We express the rate constant for Reaction 2 as a function of temperature beginning with the Arrhenius equation,

$$
k_2(T) = A \exp\left(\frac{-E_a}{RT}\right),
$$

in which E_a is the activation energy for Reaction 2, and rearrange the equation to

$$
k_2(T) = k_2(T_0) \exp\left(\frac{\alpha \Delta T}{1 + \Delta T/T_0}\right),
\tag{11.42}
$$

where T_0 is the temperature of the reactor's surroundings, $\alpha = E_a/RT_0^2$, and $\Delta T = T - T_0$. Then the rate equations are

$$
\frac{d[X]}{dt} = k_1[A] - k_2(T_0)[X] \exp\left(\frac{\alpha \Delta T}{1 + \Delta T/T_0}\right)
\tag{11.43}
$$

$$
\frac{d(\Delta T)}{dt} = \gamma[X] \exp\left(\frac{\alpha \Delta T}{1 + \Delta T/T_0}\right) - \beta \Delta T.
\tag{11.44}
$$

If A and V are the reactor's surface area and volume, ρ and c_P the density and specific heat of the reaction mixture, h the coefficient for heat transfer from the reactor to the surroundings, and $\Delta_r H$ the enthalpy for Reaction 2, the parameters β and γ in Eq. (11.44) are calculated with

$$
\beta = \frac{Ah}{V\rho c_P}
\tag{11.45}
$$

$$
\gamma = \frac{|\Delta_r H| k_2(T_0)}{\rho c_P}.
\tag{11.46}
$$

The program Thermkin integrates Eqs. (11.43) and (11.44) and plots ΔT and $[X]$ vs t.

We have seen in three of our examples that oscillating systems involving two variables, two rate equations and constant reactant concentrations have oscillations that follow a particularly simple pattern: the peaks all have the same amplitude and they occur with the same frequency. This is a *period-1 pattern*.

Autocatalytic systems described by three variables and three rate equations are considerably more complicated. The additional equation and variable make possible a seemingly endless variety of oscillation patterns. We take as an example an elaboration of the cubic autocatalator (11.33). We now recognize that Reaction 1 has an appreciable activation energy E_a, so k_1 depends on T, and that Reaction 4 is exothermic ($\Delta_r H < 0$).

The rate equations for the intermediates X and Y written in terms of the dimensionless time variable $\tau = k_4 t$, are identical to Eqs. (11.34) and (11.35), except that we now recognize the temperature dependence of k_1,

$$\frac{d[X]}{d\tau} = \frac{k_1(T)[A] - K_2[X] - k_3[X][Y]^2}{k_4} \qquad (11.47)$$

$$\frac{d[Y]}{d\tau} = \frac{k_2[X] + k_3[X][Y]^2 - k_4[Y]}{k_4}. \qquad (11.48)$$

We express $k_1(T)$ approximately in terms of the unitless temperature-related variable $\theta = (E_a)(\Delta T)/RT_0^2$, with E_a the activation energy for Reaction 1,

$$k_1(T) = k_1(T_0)e^\theta, \qquad (11.49)$$

and this is substituted in Eq. (11.47).

The third rate equation, which calculates the rate of change in θ, is similar to Eq. (11.44),

$$\frac{d\theta}{d\tau} = \gamma[Y] - \beta\theta, \qquad (11.50)$$

in which

$$\beta = \frac{Ah}{V\rho c_P k_4} \qquad (11.51)$$

$$\gamma = \frac{E_a|\Delta_r H|}{RT_0^2 \rho c_P}. \qquad (11.52)$$

The program `Cubictko` integrates Eqs. (11.47), (11.48), and (11.50) and plots [Y] vs τ. Run the program for $[A] = 0.6000$ mol L^{-1} and note that the oscillations occur in a period-1 pattern. The oscillations are more complicated when $[A] = 0.6500$ mol L^{-1}: every second peak repeats in a period-2 pattern. At $[A] = 0.6870$ mol L^{-1} there is period-4 behavior; and at $[A] = 0.6970$ mol L^{-1} period-3. At $[A] = 0.7080$ mol L^{-1}, we are confronted with *chaos*: there are oscillations but no repeat pattern.

For further study of chaos in the mathematical realm see Exercises 11-21 and 11-22.

11.6 Electrode Kinetics

The events that take place at or near an electrode during electrolysis are numerous and complicated. Reactant materials are transported in several ways to the vicinity of the electrode surface, and product materials are similarly transported away. One or more electron transfer steps, converting oxidized forms of components to reduced forms, or vice versa, take place at the electrode, and these electrode reactions may be accompanied by various nonelectrochemical reactions.

For our limited account, we will need to trim away some of these complexities to obtain a manageable model. We assume that the electrode reactions consist of two n-electron transfer steps which convert an oxidized component Ox to a reduced component Red and vice versa,

$$\text{Ox} + n\text{e}^- \underset{\text{a}}{\overset{\text{c}}{\rightleftharpoons}} \text{Red}.$$

In the forward or *cathodic* direction (c) the reaction accomplishes reduction, and in the backward or *anodic* direction (a), oxidation. We assume in our model that Ox and Red are transported to and from the electrode surface by diffusion, and that no nonelectrochemical reactions are involved. In summary, the mechanism is

$$\underset{\text{Diffusion}}{\text{Ox(b)} \rightleftharpoons} \underset{\substack{\text{Electrode}\\\text{reactions}}}{\text{Ox(s)} \underset{\text{a}}{\overset{\text{c}}{\rightleftharpoons}} \text{Red(s)}} \underset{\text{Diffusion}}{\rightleftharpoons \text{Red(b)},}$$

in which (b) and (s) represent the electroactive components Ox and Red located in the bulk of the solution and at the electrode surface.

The electrode reaction steps dominate the diffusion steps in the overall kinetics if the reaction rates are much slower than the diffusion rates. (The slow step is the bottleneck or rate-determining step.) We first assume that to be the case and develop a separate treatment of electrode reaction kinetics. We then assume the opposite situation, that the diffusion steps dominate because their rates are much less than the electrode reaction rates, and describe electrode diffusion kinetics.

Electrode Reaction Kinetics

As always in chemical kinetics, we are concerned with processes that are thermodynamically irreversible. For electrode processes an important measure of irreversibility is the *overpotential* η, defined as the difference between the reversible electrode potential E_{rev}, measured when the net electrode current density j_{net} is equal to zero, and the irreversible potential E_{irr}, measured when $j_{\text{net}} \neq 0$,

$$\eta = E_{\text{irr}} - E_{\text{rev}}. \tag{11.53}$$

Of particular interest to electrochemists is the connection between the overpotential η and the net current density j_{net}.

We approach that relationship by first writing the net rate of the electrode reaction R_{net} as the difference between the cathodic rate R_c and the anodic rate R_a,

$$R_{net} = R_c - R_a.$$

With the rates R_c and R_a expressed in terms of first-order rate laws this becomes

$$R_{net} = k_c[Ox]_b - k_a[Red]_b,$$

where k_c and k_a are cathodic and anodic rate constants, and $[Ox]_b$ and $[Red]_b$ are concentrations in the bulk of the solution.

Each occurrence of a reaction at an electrode causes n electrons to be transferred to or from the electrode. Thus R_{net} calculates the corresponding net current density j_{net}, expressed, let us say in moles of electrons per cm^2 of electrode area per second. The current density calculated in the more familiar units coulombs per cm^2 per second, or amperes per cm^2, is

$$j_{net} = nFR_{net},$$

with F the Faraday constant.

The equation that relates j_{net} and the overpotential η, known as the *Butler–Volmer equation*, is

$$j_{net} = j_0 \left[\exp\left(-\frac{n\alpha F\eta}{RT} \right) - \exp\left(\frac{n\beta F\eta}{RT} \right) \right]. \tag{11.54}$$

The parameters α and β are unitless quantities called *cathodic* and *anodic transfer coefficients*; they are restricted by $\alpha + \beta = 1$. The factor j_0 in Eq. (11.54), the *exchange current density*, expresses both the cathodic and anodic currents when the electrode operates reversibly (with $j_{net} = 0$).

Over part of the range of η values the Butler–Volmer equation simplifies to a single exponential term because the other exponential term is negligible. Thus for η large in magnitude and positive

$$|j_{net}| \cong j_0 \exp\left(\frac{n\beta F\eta}{RT} \right),$$

or

$$\ln|j_{net}| \cong \ln j_0 \frac{n\beta F\eta}{RT}. \tag{11.55}$$

For η large in magnitude and negative

$$\ln|j_{net}| \cong \ln j_0 - \frac{n\alpha F\eta}{RT}. \tag{11.56}$$

In both cases a plot of the linear form

$$\ln|j_{net}| = a + b\eta, \tag{11.57}$$

called a *Tafel plot*, is indicated.

The program `Butler` displays plots of the Butler–Volmer equation and Tafel plots. Run the program and note the effects of changing the parameters α, β and j_0.

Electrode Diffusion Kinetics

We now look at an entirely different kind of electrochemistry based on electrode processes dominated by diffusion rates rather than electrode reaction rates. We assume that the electrode reaction is rapid and reversible, and that the diffusion rates are relatively much slower.

Picture a *diffusion layer* between the bulk of the solution and the electrode surface within which the concentration of Ox in our model scheme falls from $[Ox]_b$, the concentration of Ox in the bulk of the solution, to $[Ox]_s$, the concentration at the electrode surface. The $[Ox]$ gradient across this layer is approximately $([Ox]_s - [Ox]_b)/x_{Ox}$, where x_{Ox} is the width of the Ox diffusion layer. The rate of diffusion of Ox across the diffusion layer toward the electrode is proportional to this gradient and the proportionality factor is the diffusion coefficient D_{Ox} for Ox,

$$\text{Rate} = \frac{-D_{Ox}([Ox]_s - [Ox]_b)}{x_{Ox}}.$$

The corresponding diffusion current density is

$$j = (nF)(\text{Rate}) = \frac{nFD_{Ox}([Ox]_b - [Ox]_s)}{x_{Ox}}. \qquad (11.58)$$

As the potential E applied to the electrode is made more negative the rate of reduction of Ox to Red increases and that decreases $[Ox]_s$, finally to $[Ox]_s = 0$, where the diffusion current has its limiting value j_L,

$$j_L = \frac{nFD_{Ox}[Ox]_b}{x_{Ox}}, \qquad (11.59)$$

or

$$[Ox]_b = \frac{j_L x_{Ox}}{nFD_{Ox}}. \qquad (11.60)$$

The magnitude of the current density is also determined by diffusion of Red away from the electrode,

$$j = \frac{nFD_{Red}([Red]_s - [Red]_b)}{x_{Red}}.$$

We consider the case of no Red in the bulk of the solution, so $[Red]_b = 0$, and

$$[Red]_s = \frac{x_{Red} j}{nFD_{Red}}. \qquad (11.61)$$

The electrode reaction we are concerned with is rapid and thermodynamically reversible. We can therefore assume that the Nernst equation applies to the reaction at the electrode surface where the concentrations of the electroactive components are $[Ox]_s$ and $[Red]_s$,

$$E = E^{0\prime} - \frac{RT}{nF} \ln \frac{[Red]_s}{[Ox]_s}, \tag{11.62}$$

in which $E^{0\prime}$ is a standard electrode potential measured for specific conditions.

Combining Eqs. (11.60), (11.61), and (11.62) and rearranging we have

$$\frac{j}{j_L} = \frac{1}{1 + \exp\left[\dfrac{nF(E - E_{1/2})}{RT}\right]}, \tag{11.63}$$

where $E_{1/2}$, the *half-wave potential*, is the value of the applied potential E which gives j its half value $j_L/2$.

Equation (11.63) is not as simple as it looks, because the two diffusion layers for Ox and Red do not have the same widths, so x_{Ox} and x_{Red} depend not only on the diffusion coefficients D_{Ox} and D_{Red} but also on the time variable t. As a result, the half-wave potential $E_{1/2}$ and the limiting current j_L are also time dependent: what you get in an electrode measurement may depend on when you make the measurement.

One way to cope with this complication is to make each measurement at a certain well-controlled time; that is the tactic used in the electrode design called the *dropping-mercury electrode*. Another approach is to introduce efficient stirring of the solution (until now we have assumed that the solution is unstirred), so that the diffusion layers have time-independent, steady-state structures; this is accomplished in the *rotating-disk electrode*.

The program Dme simulates a polarogram (a plot of the electrode current i vs the applied potential E) taken with a dropping-mercury electrode. The plot has "teeth," each of which follows the growth of a drop until the drop falls. The top of each tooth marks the point where the drop falls. The teeth maxima are easy to locate and together they define the shape of the polarogram. The cycle of events at the dropping-mercury electrode is well enough defined to allow calculation of the limiting current i_L according to

$$i_L = 706nD_{Ox}^{1/2}[Ox]_b u^{2/3} t^{1/6}, \tag{11.64}$$

written for the electroactive component Ox. In this equation D_{Ox} is a diffusion coefficient measured in $m^2\ s^{-1}$, $[Ox]_b$ is a concentration in $mol\ m^{-3}$, u is the mass flow rate of mercury in $kg\ s^{-1}$ and t is time in the life of a drop in s. Run the program Dme and note that it simulates the polarogram obtained with two electroactive components Ox_1 and Ox_2 in the solution; their half-wave potentials are -0.6 and -0.8 V.

The program Rme simulates a voltammetry plot obtained with a rotating-disk electrode for the same two electroactive components. Diffusion in this case takes place across a thin stagnant layer next to the electrode surface. The physical status of the stagnant layer has been treated theoretically and an equation for the limiting current derived,

$$i_L = 0.620 nFAD_{Ox}^{2/3}[Ox]_b v^{-1/16} \omega^{1/2}, \tag{11.65}$$

where F is Faraday's constant, A is the area of the electrode surface measured in m^2, D_{Ox} is a diffusion coefficient in $m^2 s^{-1}$, $[Ox]_b$ is a concentration in $mol\,m^{-3}$, v is the kinematic viscosity $(= \eta/d$, where η is the viscosity in $kg\,m^{-1}\,s^{-1}$ and d is the density in $kg\,m^{-3})$, and ω is the angular rotational speed of the electrode in radians $s^{-1}[= (2\pi)(rpm)/60]$.

11.7   Stochastic Kinetics

Chemical kinetics, like chemical thermodynamics, is designed to describe macroscopic events. The rate equations of chemical kinetics are not accurate when a small number of molecules is involved ("small" in this case means less than about 10000). In this section we take a completely different approach to kinetics and simulate the kinetic data we would obtain if we followed chemical rate processes in small systems.

The new method is called *stochastic kinetics*, and it is based on a probability analysis. To illustrate the strategy, we consider the reaction scheme

$$A \xrightarrow{1} X$$
$$X \xrightarrow{2} R$$
$$R \xrightarrow{3} X$$
$$A \xrightarrow{4} S.$$

Suppose that at some time t the numbers of A, X and R molecules are n_A, n_X, and n_R, so rates of the reactions are

$$r(1) = k_1 n_A$$
$$r(2) = k_2 n_X$$
$$r(3) = k_3 n_R$$
$$r(4) = k_4 n_A,$$

where the k's are rate constants. Then the probability $p(1)$ for Reaction 1 is

$$p(1) = \frac{r(1)}{r_{sum}},$$

in which $r_{sum} = \sum_i r(i)$, and similarly for the other reactions.

In the stochastic simulation, we allow the reactions to take place randomly as dictated by the probabilities $p(1)$, $p(2)$, $p(3)$, and $p(4)$. We might, for example, have $p(1) = 0.1$, $p(2) = 0.2$, $p(3) = 0.4$, and $p(4) = 0.3$ at some time t. This tells us that Reactions 2, 3, and 4 are 2, 4, and 3 times as probable as Reaction 1. The simulation algorithm makes use of running sums of these probabilities,

$$p(1) = 0.1$$

$$p(1) + p(2) = 0.3$$

$$p(1) + p(2) + p(3) = 0.7$$

$$p(1) + p(2) + p(3) + p(4) = 1.0.$$

To select a reaction, the algorithm picks a random number p_1 from a uniform distribution between 0 and 1, and then makes a reaction decision based on the range within which p_1 lies:

With p_1 in the range,	the reaction selected is:
0.0–0.1	1
0.1–0.3	2
0.3–0.7	3
0.7–1.0	4

This accomplishes the desired result in our example: Reactions 2, 3, and 4 are 2, 4, and 3 times as probable as Reaction 1.

After a reaction has been selected the numbers n_A, n_X, and n_R are changed according to the reaction's stoichiometry. For the example the rules are:

Reaction	Δn_A	Δn_X	Δn_R
1	−1	+1	0
2	0	−1	+1
3	0	+1	−1
4	−1	0	0

The next step in the algorithm is to calculate the time τ before another reaction (any reaction) occurs. This is done by picking another random number p_2 from a uniform distribution in the range 0 to 1, and calculating τ according to

$$\tau = \frac{\ln(1/p_2)}{r_{sum}}, \tag{11.66}$$

which distributes the τ's according to $r_{sum}e^{-r_{sum}\tau}$, making shorter τ's more probable than longer ones.

The algorithm then increments the time t by τ, and goes through another cycle of calculations: the $r(i)$'s and $p(i)$'s are recalculated, another pair of random numbers p_1 and p_2 is generated, a reaction decision is made, n_A, n_X, and n_R are changed, τ is recalculated, t is incremented by τ, and so forth. For more details on this method see Gillespie's paper.

The program Stokin implements this calculation for the reaction scheme

$$A \xrightarrow{1} X$$
$$X \xrightarrow{2} A$$
$$X \xrightarrow{3} R,$$

assuming that the system contains 1000 molecules (of A, X, or R), and plots n_X vs t. Stokin also calculates and plots n_X deterministically, that is, using the rate equations

$$\frac{d[A]}{dt} = -k_1[A] + k_2[X]$$

$$\frac{d[X]}{dt} = k_1[A] - k_2[X] - k_3[X].$$

Comparison with the stochastic calculation shows the limitations of the deterministic calculation for small systems.

11.8   Exercises

Reactions in Beams

11-1   The reaction $CH_3I + K \rightarrow KI + CH_3$ has been studied in molecular beams by Rulis and Bernstein. In a series of measurements, the average speed of K in one beam was 601 m s^{-1} and of CH_3I in the other was 256 m s^{-1}. The maximum flux of the product KI was obtained at a laboratory scattering angle of $\Theta = 85°$ and the KI speed there was about 450 m s^{-1}. Calculate the center-of-mass scattering angle θ and the relative precollision and postcollision energies E_t and E_t'. Assume that the methyl group CH_3- behaves as if it were an atom.

11-2   In the text, we developed methods for calculating precollision and postcollision relative translational kinetic energies E_t and E_t'. E_t is part of the energy input that initiates a reaction $A + BC \rightarrow AB + C$. Other important energy inputs are the rotational and vibrational energy of BC, $E_r(BC)$, and $E_v(BC)$, and the energy of the reaction itself. The latter is the difference between the dissociation energies $D_0(AB)$ and $D_0(BC)$ for AB and BC, because the reaction is the chemical sum of the dissociation of BC, $BC \rightarrow B + C$, and the reverse of the dissociation of AB, $A + B \rightarrow AB$. The total energy input to the reactive collision is therefore

$$E_t + E_r(BC) + E_v(BC) + D_0(AB) - D_0(BC).$$

For energy conservation, this precollision total energy must be balanced by the postcollision total energy comprising the relative translational kinetic energy E'_t and the "internal" energy $E_i(AB)$ of the product, that is, its combined rotational and vibrational energy. Thus the full energy conservation statement is

$$E_t + E_r(BC) + E_v(BC) + D_0(AB) - D_0(BC) = E'_t + E_i(AB). \quad (11.67)$$

The crossed-beam experiment provides information on E_t and E'_t. Other quantities on the left side of Eq. (11.67) are usually obtainable from spectroscopic data, so the equation is usefully solved for $E_i(AB)$,

$$E_i(AB) = E_t - E'_t + E_r(BC + E_v(BC) + D_0(AB) - D_0(BC). \quad (11.68)$$

Consider again the reaction $CH_3I + K \to KI + CH_3$ mentioned in Exercise 11-1. For the experiment described initial rotational and vibrational energies for CH_3I and relevant dissociation energies are

$$E_r(CH_3I) = 3.6 \text{ kJ mol}^{-1}$$

$$E_v(CH_3I) = 0.9 \text{ kJ mol}^{-1}$$

$$D_0(CH_3I) = 226.0 \text{ kJ mol}^{-1}$$

$$D_0(KI) = 326.4 \text{ kJ mol}^{-1}.$$

Estimate the energy allocated by the reaction to the CH_3I internal modes of motion.

11-3   Consider again the reaction $K + I_2 \to KI + I$ mentioned in Example 11-1. Use Eq. (11.68), derived in the last exercise, to calculate the energy allocated by the reaction to the KI internal modes of motion. For the experiment described initial rotational and vibrational energies and relevant dissociation energies are

$$E_r(I_2) = 2.8 \text{ kJ mol}^{-1}$$
$$E_v(I_2) = 1.8 \text{ kJ mol}^{-1}$$
$$D_0(I_2) = 148.5 \text{ kJ mol}^{-1}$$
$$D_0(KI) = 318.0 \text{ kJ mol}^{-1}.$$

11-4   The calculation done in Example 11-1 determines the center-of-mass scattering angle θ corresponding to the laboratory scattering angle $\Theta = 30°$ where the maximum flux of the product KI is observed in a molecular beam experiment. Offhand, you might expect this angle would also locate the maximum flux in the center-of-mass system. But the connection between the laboratory flux $j(\text{lab})$ and the center-of-mass flux $j(\text{cm})$ is more complicated than that: for a reaction $A + BC \to AB + C$,

$$j(\text{cm}) = \left(\frac{u'_{AB}}{v'_{AB}}\right)^2 j(\text{lab}). \quad (11.69)$$

Write a modification of the program Newton to include this calculation. Use the data quoted below for observed values of $j(\text{lab})$ at various laboratory scattering angles Θ to calculate corresponding values of $j(\text{cm})$ at various center-of-mass scattering angles θ. At what value of θ does $j(\text{cm})$ have a maximum value?

$\Theta/\text{degrees}$	$v'_{\text{KI}}/(\text{m s}^{-1})$	$j(\text{lab})/(\text{arb. units})$
-15	480	3.5
-5	500	3.7
10	450	8.7
20	390	9.3
25	380	9.5
30	360	10.0
40	340	10.0
50	350	8.5
60	350	5.5
70	380	2.7
80	500	2.5
100	510	2.5

Source: Gillen, Rulis, and Bernstein, 1971.

Potential Energy Surfaces

11-5   Run the program Leps with

$$D_e(\text{AB}) = 600 \text{ kJ mol}^{-1}$$

$$D_e(\text{BC}) = 200 \text{ kJ mol}^{-1}$$

$$D_e(\text{AC}) = 200 \text{ kJ mol}^{-1}$$

$$K = -0.1,$$

and other parameters unchanged. Estimate the activation energy indicated by the potential energy surface plotted.

11-6   Use the program Leps to plot a potential energy surface for a reaction with an endoergicity of about 250 kJ mol^{-1} and an activation energy of about 400 kJ mol^{-1}.

11-7   Revise the program Leps so it accomodates a (fixed) bent geometry for the triatomic transition states rather than the linear geometry assumed in Leps.

11-8   Write a program which plots a potential energy profile for a reaction, that is, a plot of potential energy versus distance along the reaction path. Locate the coordinates for points on the path by sketching the path on a contour map of the potential energy surface made by Leps, and then locating equidistant points separated by 0.25 Å along the path. Also have the program plot a potential energy for the point whose coordinates locate the activated complex.

Activated Complex Theory

11-9   Schatz and Walch did an ab initio calculation of the potential energy surface for the reaction

$$OH + H_2 \longrightarrow H_2O + OH.$$

They calculated 25940 $J\,mol^{-1}$ for the reaction's threshold energy. Other results of theirs have been incorporated in the data included in the file Chap11.m for the activated complex (designated OHHHac in Chap11.m). Use the program Eyring2 to calculate and plot rate constants for the reaction in the temperature range 300 to 1000 K. Also plot the empirical equation,

$$k/(L\,mol^{-1}\,s^{-1}) = (8.02 \times 10^8)\left(\frac{T}{298\ K}\right)^{1.73} \exp\left(-\frac{1605.7\ K}{T}\right),$$

obtained from the NIST kinetics database.

11-10   Use the program Eyring2 to calculate a rate constant for the reaction

$$H + I_2 \longrightarrow HI + I,$$

at 300 K. Data are included in the file Chap11.m for the activated complex (called HIIac in Chap11.m). Assume that the reaction has no threshold energy. An observed value for the rate constant is $3.7 \times 10^{11}\ L\,mol^{-1}\,s^{-1}$.

11-11   The gaseous component HNC, hydrogen isocyanide, isomerizes to hydrogen cyanide, HCN, in a unimolecular reaction, HNC → HCN. This reaction has apparently not been studied experimentally because HNC is difficult to prepare in the laboratory (the molecule is found in interstellar space, however). Nevertheless, activated complex theory can be used to estimate rate constants for the reaction at any temperature. Use the program Eyring1 to calculate rate constants for the isomerization reaction in the temperature range 300 to 1000 K, with 134 $kJ\,mol^{-1}$ for the reaction's threshold energy.

11-12   In the last exercise you calculated rate constants for the isomerization reaction HNC → HCN. Consider now the reverse isomerization reaction HCN → HNC. Use the program Eyring1 to calculate rate constants for this reaction in the temperature range 300 to 1000 K. Note that one calculation gave 62 $kJ\,mol^{-1}$ as the exoergicity of the forward reaction HNC → HCN. Compare your calculated values of the rate constants with values calculated using the Arrhenius parameters $A = 3.5 \times 10^{13}$ s^{-1} and $E_a = 197$ kJ mol^{-1}, measured by Lin et al.

Unimolecular Reactions

11-13 Run the program Beyer with the data supplied in the program. (You may want to schedule another activity while you wait for the program to finish its task; the calculation is very slow.)

11-14 The two programs Beyer and Whitten do the same calculation, the former with direct-count algorithms, and the latter with analytical procedures. Run Whitten and check its result against that obtained with Beyer in Exercise 11-13. Restrict wMax in Whitten to 0.8 wz.

11-15 The data file c2h5cl.dat contains densities of states $N(E)$ and sums of states $G^{\ddagger}(E - E_0)$ for the activated complex in the unimolecular decomposition reaction,

$$C_2H_5Cl \longrightarrow C_2H_4 + HCl,$$

at 1000 K and 750 Torr. Use this file and the program Rrkm to calculate and plot microscopic rate constants for the above reaction.

Oscillating Reactions

11-16 Prove that the following reaction scheme, involving a quadratic autocatalysis step,

$$A \xrightarrow{1} Y$$
$$B + X \xrightarrow{2} R + Y$$
$$Y + X \xrightarrow{3} 2X$$
$$X \xrightarrow{4} S,$$

has no unstable steady states.

11-17 Consider the following autocatalytic scheme,

$$A + X \xrightarrow{1} 2X$$
$$X + Y \xrightarrow{2} 2Y$$
$$Y \xrightarrow{3} R,$$

known as the *Lotka–Volterra mechanism*, which cannot have stable steady states under any conditions. Revise the program Brussels so it calculates and plots [X] and [Y] for this scheme with $[A] = 1$ mol L^{-1}, $k_1 = 1$ L mol^{-1} s^{-1}, $k_2 = 0.5$ L mol^{-1} s^{-1}, $k_3 = 0.8$ s^{-1}, and $[X]_0 = [Y]_0 = 1$ mol L^{-1}.

11-18   In this exercise you can derive the rate equations for the thermo-chemical oscillator, Scheme (11.41). Express the rate constant for Reaction 2 as a function of temperature beginning with the Arrhenius equation $k_2(T) = Ae^{-E_a/RT}$, in which E_a is the activation energy for Reaction 2. Rearrange this to

$$k_2(T) = k_2(T_0) \exp\left(\frac{\alpha \Delta T}{1 + \alpha \Delta T / T_0}\right), \qquad (11.70)$$

in which $\alpha = E_a/RT_0^2$. Use this result to derive a rate equation for X. Derive an equation that expresses $d\Delta T/dt$, the rate of change in ΔT, by writing an enthalpy balance statement for the reactor in which the reaction takes place,

$$\begin{pmatrix} \text{Net rate of} \\ \text{enthalpy change} \end{pmatrix} = \begin{pmatrix} \text{Rate of} \\ \text{enthalpy input} \end{pmatrix} - \begin{pmatrix} \text{Rate of} \\ \text{enthalpy output} \end{pmatrix},$$

with

$$\begin{pmatrix} \text{Net rate of} \\ \text{enthalpy change} \end{pmatrix} = (V\rho c_P)\frac{dT}{dt} = (V\rho c_P)\frac{d\Delta T}{dt}$$

$$\begin{pmatrix} \text{Rate of} \\ \text{enthalpy input} \end{pmatrix} = |\Delta_r H| V k_2(T)[\text{X}]$$

$$\begin{pmatrix} \text{Rate of} \\ \text{enthalpy output} \end{pmatrix} = (Ah)(\Delta T),$$

where the parameters A and V are the reactor's surface area and volume, ρ and c_P are the density and specific heat of the reaction mixture, h is the co-efficient of heat transfer from the reactor to the surroundings, and $\Delta_r H$ is the enthalpy for Reaction 2. Combine these equations to obtain Eq. (11.44).

11-19   The program `Cubecat` does its calculation for Scheme (11.33) as-suming that the concentration [A] of A is constant. Modify the program so changes in [A] are allowed. Have the program plot [A] and 10[X] from $\tau = 0$ to 1000 s. Use $[A]_0 = 0.5$ mol L^{-1}.

11-20   The program `Thermkin` does its calculation for Scheme (11.41) as-suming that the concentration [A] of A is constant. Modify the program so changes in [A] are allowed. Have the program plot ΔT, [A] and 10[X] from $t = 0$ to 20 s. Use $[A]_0 = 1.5$ mol L^{-1}.

11-21   You can make a study in this exercise of the astonishingly complex periodic and aperiodic (chaotic) behavior of the simple recursion relation

$$x(n+1) = Ax(n)[1 - x(n)] \qquad (11.71)$$

for various values of the parameter A in the range 3 to 4. Begin the calcu-lation with $x(0) = 0.5$, then use the Eq. (11.71) to calculate $x(1)$ from $x(0)$, then $x(2)$ from $x(1)$, etc. A series of numbers calculated this way may or may not show periodic behavior. For example, with $A = 3.5$, Eq. (11.71) shows period-4 behavior, but with $A = 3.75$, aperiodic or chaotic behavior. Write a program that calculates an extended series of numbers from Eq. (11.71) and searches for periodic behavior. If it succeeds, have the program print out a typical cycle of numbers.

11-22 Chaos theory introduces a parameter

$$\lambda = \lim_{n\to\infty} \frac{1}{n} \sum_{i=0}^{n} \log_2 \left|\frac{\partial f}{\partial x}\right|, \qquad (11.72)$$

called the *Lyapounov exponent*, to test for periodic or aperiodic (chaotic) behavior. A negative value of λ signifies periodic behavior and a positive value aperiodic behavior. The function f is the right side of the equation of interest. For example, in Eq. (11.71), $f = Ax(i)[1 - x(i)]$, and

$$\left|\frac{\partial f}{\partial x}\right| = |A[1 - 2x(i)]|,$$

so

$$\log_2 \left|\frac{\partial f}{\partial x}\right| = \frac{\ln |A[1 - 2x(i)]|}{\ln 2}.$$

Write a program which calculates λ from Eq. (11.71) for a given value of the parameter A. Assume that $n = 100$ is an "infinite" value of n in Eq. (11.72).

Electrode Kinetics

11-23 Run the program Butler and note the effects of changing the parameters n, α, and β.

11-24 Run the program Dme with the half-wave potentials -0.6 and -0.7 V for the electroactive components. What problem does this simulation illustrate?

11-25 The programs Dme and Rde simulate voltammetry plots for electrochemical systems dominated by electrode reaction kinetics. Electrode reactions are assumed to be slow enough relative to diffusion rates so there are no concentration gradients across the diffusion layers. If this assumption is not valid the Butler–Volmer equation has to be modified, perhaps to

$$j = \frac{j_0 j_L \left[\exp\left(-\frac{\alpha Fn}{RT}\right) - \exp\left(\frac{\beta Fn}{RT}\right)\right]}{j_L + j_0 \left[\exp\left(-\frac{\alpha Fn}{RT}\right) - \exp\left(\frac{\beta Fn}{RT}\right)\right]}, \qquad (11.73)$$

where j_L is the limiting current. Rewrite the program Dme so the current density is calculated this way. Assume that $j_L = 0.01$ A cm^{-2} for $\eta < 0$ and -0.01 A cm^{-2} for $\eta > 0$. Leave other parameters in Dme unchanged.

Stochastic Kinetics

11-26 The program Stokin is designed to make calculations for reaction schemes comprising more than one reaction. The program can, however, be simplified so it handles a single reaction. Make this adaptation and have the revised program calculate and plot n_A and n_B stochastically and deterministically for the bimolecular reaction

$$A + B \longrightarrow R + S.$$

Use $k = 0.0001$ s^{-1} for the rate constant and $n_{A0} = 500$, $n_{B0} = 1000$ for initial numbers of A and B molecules. Have the program make a calculation for the time t in the range 0 to 50 s.

11-27 Adapt the program Stokin so it calculates numbers of molecules of X formed as an intermediate in the sequence of reactions,

$$A \xrightarrow{1} X$$
$$X \xrightarrow{2} R.$$

Use $k_1 = 0.1$ s^{-1} and $k_2 = 0.2$ s^{-1} for the rate constants, and $n_{A0} = 1000$ and $n_{X0} = 0$ for the initial numbers of A and X molecules. Make the calculation for t in the range 0 to 50 s.

11-28 Adapt the program Stokin so it calculates and plots numbers of molecules of X formed as an intermediate in the scheme

$$A \xrightarrow{1} X$$
$$A \xrightarrow{2} R$$
$$X \xrightarrow{3} R$$
$$R \xrightarrow{4} X.$$

Use $k_1 = 0.1$ s^{-1}, $k_2 = 0.2$ s^{-1}, $k_3 = 0.3$ s^{-1} and $k_4 = 0.4$ s^{-1}, and $n_{A0} = 100$, $n_{X0} = 0$, and $n_{R0} = 0$ for the initial numbers of molecules. Make the calculation for t in the range 0 to 50 s.

Appendix A
A List of the Program and Data Files on the Disk*

`2dnmr`	Simulates a ^{1}H COSY two-dimensional NMR plot (5.5).
`Abc`	Calculates principal moments of inertia for linear and non-linear molecules (4.5).
`Acetate`	Calculates the pH in an acetate buffer (3.2).
`Anderko`	Solves the Anderko–Pitzer nonideal gas law (1.2).
`Aorbital`	Calculates and plots H atomic orbitals in two dimensions (4.3).
`Atp1`	Calculates and plots the apparent standard Gibbs energy $\Delta_r G^{0\prime}$ for the ATP hydrolysis reaction at various pMg's and a specified pH and ionic strength (3.3).
`Atp2`	Calculates and plots the apparent standard Gibbs energy $\Delta_r G^{0\prime}$ for the ATP hydrolysis reaction at various pH's and a specified pMg and ionic strength (3.3).
`Beattie`	Solves the Beattie–Bridgeman nonideal gas law (1.2).
`Beyer`	Calculates and plots the density of states $N(E)$ and the sum of states $G(E)$ for a molecule's vibrational modes using the Beyer–Swinehart direct-count algorithm (11.4).
`Biochem`	Calculates and plots the apparent standard Gibbs energy $\Delta_r G^{0\prime}$ for the F $\rightarrow$ M biochemical reaction at various pH's (3.3).
`Biokin`	Calculates and plots extracellular glucose and insulin concentrations as functions of time (10.4).
`Branch`	Calculates and plots H, O, OH and O_2 concentrations for the branching catalytic cycle that drives the $2H_2 + O_2 \rightarrow 2H_2O$ reaction (10.2).
`Brussels`	Calculates and plots concentrations of intermediates in the autocatalytic Brusselator model (11.5).
`Butler`	Calculates and plots the Butler–Volmer and Tafel equations relating electrode current and overpotential (11.6).

*Mathematica program files on the Disk have the extension `.ma` for Version 2.2, and `.nb` for Version 3, QuickBASIC program files the extension .BAS, and data files the extension `.dat` or `.m`. Numbers in parentheses refer to section numbers.

c2h5	Contains $N(E)$ and $G^{\ddagger}(E)$ data for the $C_2H_5 \rightarrow C_2H_4 + H$ reaction (11.4).
c2h5cl	Contains $N(E)$ and $G^{\ddagger}(E)$ data for the $C_2H_5Cl \rightarrow C_2H_5 + HCl$ reaction (11.4).
Catcycle	Integrates and plots the rate equations for a catalytic cycle (10.2).
Chain	Integrates and plots rate equations for chain polymerization reactions (10.3).
Chap1	Data file for Chapter 1.
Chap2	Data file for Chapter 2.
Chap3	Data file for Chapter 3.
Chap4	Data file for Chapter 4.
Chap8	Data file for Chapter 8.
Chap11	Data file for Chapter 11.
Chase	Calculates accurate values for standard molar entropies, molar enthalpies, molar heat capacities, and chemical potentials for gases whose molecules are diatomic (8.4).
Chemkin	Calculates and plots changes in concentrations of components in the bimolecular reversible reaction $A + B \rightleftarrows R + S$ (10.1).
Coal	Calculates equilibrium partial pressures for the coal gasification reactions (3.2).
Coil1	Calculates and plots in three dimensions a random chain, either freely rotated or freely jointed (7.1).
Coil2	Calculates a mean value of the end-to-end distance for freely rotated random chains (7.1).
Coil3	Calculates a mean value of the end-to-end distance for freely jointed random chains (7.1).
Collide	Solves the classical equations of motion for two colliding molecules and plots the trajectory of one of the molecules with respect to the center of mass of the two-molecule system (9.2).
Cpd2	Calculates and plots D_2 standard spectroscopic molar heat capacities for pure ortho D_2, pure para D_2, the "normal" 2/1 ortho-para mixture, and the equilibrium mixture (8.6).
Cph2	Calculates and plots H_2 standard spectroscopic molar heat capacities for pure ortho H_2, pure para H_2, the "normal" 3/1 ortho-para mixture, and the equilibrium mixture (8.6).
Cphd	Calculates and plots standard spectroscopic molar heat capacities for HD (8.6).
Cubecat	Calculates and plots changes in concentrations of components participating in the isothermal cubic catalator (11.5).
Cubictko	Calculates and plots changes in concentrations of components participating in the nonisothermal cubic autocatalator (11.5).
Cycle	Calculates and plots changes in concentrations of components participating in a simple cycle of reactions (10.2).

Debye	Calculates the Debye entropy extrapolation from a low temperature to 0 K (2.3).
Deltag1	Calculates standard reaction enthalpies, entropies, and Gibbs energies at high temperatures using a polynomial heat capacity formula (2.1).
Deltag2	Calculates standard reaction enthalpies, entropies and Gibbs energies at high temperatures using the Maier–Kelley heat capacity formula (2.1).
Deltag3	Calculates standard reaction Gibbs energies at high temperatures using a special empirical formula (2.1).
Diffuse1	Simulates in an animated plot diffusion from a thin plane source (9.3).
Diffuse2	Simulates in an animated plot diffusion from a step-function source (9.3).
Dme	Simulates a polarogram taken by a dropping-mercury electrode (11.6).
Duality	Simulates two-slit, interference-diffraction patterns produced by an electron beam (4.3).
Elecdiff	Interprets electron diffraction data for $SiCl_4(g)$ (6.2).
Entropy	Calculates heat capacity integrals for the determination of calorimetric entropies (2.3).
Error	Calculates uncertainty in a calculation propagated by uncertainties in the independent variables (1.1).
Esr	Calculates and plots a first-order ESR multiplet involving spin-1/2 nuclei, and draws the corresponding inverted tree that interprets the multiplet (5.5).
Eyring1	Uses activated complex theory to calculate approximate rate constants for unimolecular reactions (11.3).
Eyring2	Uses activated complex theory to calculate approximate rate constants for bimolecular reactions (11.3).
fe2sio4	File containing powder x-ray diffraction data for $Fe_2SiO_4(s)$ (fayalite) (6.1).
Fermi	Calculates the Fermi level and carrier concentrations for semiconductors (6.3).
Gibbs	Calculates and plots $G(\xi)$ and $\Delta_r G = (\partial G/\partial \xi)_{P,T}$ for the reaction $A(g) \rightarrow R(g) + S(g)$ (2.8).
Gouy1	Calculates and plots surface potentials for various surface charge densities according to the Gouy–Chapman theory of the electric double layer (6.6).
Gouy2	Calculates and plots potentials at various distances from a charged surface according to the Gouy–Chapman theory of the electric double layer (6.6).
Gouy3	Calculates and plots concentrations of anions and cations at various distances from a charged surface according to the Gouy–Chapman theory of the electric double layer (6.6).
Gplot	Calculates and plots a surface representing Gibbs energy de-

	partures for a gas at pressures and temperatures above the critical point (2.2).
Gprofile	Plots a Gibbs energy profile for the biochemical glycolysis scheme of reactions which converts glucose to lactate (3.3).
Haber	Calculates equilibrium partial pressures for the ammonia-synthesis reaction (3.2).
Hartree	Calculates Hartree–Fock molecular orbitals for LiH using H1s, Li1s, Li2s, and Li2p atomic orbitals (4.7).
hcl	File containing experimental data for the rotational–vibrational spectrum of HCl(g) (5.2).
Henry	Compares real and Henry's-law behavior for a component in a binary mixture (2.6).
Hodgkin	Uses the Hodgkin–Huxley equations to calculate squid axon action potentials for a voltage-clamp situation (10.5).
Hplot	Calculates and plots a surface representing enthalpy departures for a gas at pressures and temperatures above the critical point (2.2).
Hueckel	Calculates Hückel (or Hueckel) orbital energies and orbital coefficients for π-electron systems (4.7).
Irft	Calculates and plots a simulated interferogram and then calculates its Fourier transform (5.6).
Keenan1	Calculates steam densities according to the method of Keenan, Keyes, Hill, and Moore (1.2).
Keenan2	Calculates specific internal energies, enthalpies and entropies of steam according to the method of Keenan, Keyes, Hill, and Moore (2.2).
Krebs	Integrates and plots rate equations for a cycle of reactions that has features in common with the biochemical Krebs cycle (10.2).
Lambda	Calculates limiting molar conductivities using an extension of the Onsager–Debye–Hückel equation (9.4).
Leed	Calculates and plots substrate and surface, direct, and reciprocal lattices for crystalline surface phases (6.7).
Leps	Calculates and plots contour maps and three-dimensional representations of potential energy surfaces for three-atom reactions according to the London–Eyring–Polanyi–Sato method (11.2).
Limcycle	Demonstrates a limit cycle for the Brusselator model in a parametric plot of [X] vs [Y] at various times (11.5).
Linreg	Calculates coefficients for the best fit of data pairs to a fitting function in which the coefficients occur linearly (1.3).
Maxwell1	Calculates and plots the Maxwell distribution of molecular speeds at various temperatures (9.1).
Maxwell2	Calculates and plots the Maxwell distribution of molecular speeds at various molar masses. (9.1).

MC1	Uses a Monte Carlo simulation to calculate a radial distribution function for a "hard-sphere" liquid in two dimensions; a QuickBASIC program (6.5).
MC2	Uses a Monte Carlo simulation to calculate a radial distribution function for a "Lennard-Jones" liquid in two dimensions; a QuickBASIC program (6.5).
MD1	A molecular-dynamics simulation of a "hard-sphere" liquid in three dimensions; a QuickBASIC program (6.4).
MD2	A molecular-dynamics simulation of a "hard-sphere" liquid in two dimensions; a QuickBASIC program (6.4).
MD3	Calculates radial distribution functions from two-dimensional data obtained in a molecular dynamics simulation of a "hard-sphere" liquid; a QuickBASIC program (6.4).
Mixing	Calculates and plots changes in statistical entropies for mixing of molecules of two different kinds (8.1).
Morbital	Calculates and plots H_2 molecular orbitals (4.3).
Morse	Calculates and plots potential energy curves according to the Morse function for electronic states of diatomic molecules and superimposes vibrational energy levels plotted between the classical turning points (4.5).
Newton	Calculates collision kinetic energies and scattering angles for crossed molecular beam experiments and plots a Newton diagram (11.1).
Nmr	Calculates and plots first-order NMR multiplets involving spin-1/2 nuclei, and draws the inverted trees that interpret the multiplets (5.5).
Nmrft	Calculates and plots a simulated FID signal, and then calculates and plots its Fourier transform (5.6).
Onsager	Calculates ionization constants of weak electrolytes from molar conductivity data using the Onsager and Debye–Hückel equations (9.4).
Oregon	Calculates and plots concentrations of components participating in the Oregonator model (11.5).
Pattersn	Plots points and lines for idealized Patterson maps (6.1).
Peaks	Locates peak maxima in a list of spectral data (5.2).
Peng1	Solves the Peng–Robinson nonideal gas law (1.2).
Peng2	Calculates molar volumes, compressibility factors, molar enthalpy departures, and molar entropy departures for nonideal gases using the Peng–Robinson nonideal gas law (2.2).
Peng3	Calculates molar volumes, compressibility factors and fugacity coefficients for nonideal gases using the Peng–Robinson nonideal gas law (2.2).
Perrin1	Calculates and plots the frictional ratio $f_r = f/f_0$ as a function of the axial ratio $p = b/a$ for oblate and prolate ellipsoids of revolution (7.3).

Perrin2	Calculates the axial ratio $p = b/a$ given the frictional ratio $f_r = f/f_0$, or vice versa, for oblate and prolate ellipsoids of revolution (7.3).
Pitzer	Calculates mean activity coefficients for strong electrolytes according to Pitzer's equation (2.7).
PIXEL	Creates the file graph.dat read by PLOTAXES when it plots axes; a QuickBASIC program.
PLOTAXES	A QuickBASIC submodule which plots axes, a frame and tickmarks; and prints labels for the tickmarks and titles for the axes.
Powder	Indexes x-ray reflections obtained from powder samples (6.1).
Powell	Calculates $\Delta_r G$ for metamorphic geochemical reactions involving the gaseous components CO_2 and H_2O (3.1).
Raoult	Compares real and Raoult's-law behavior for a component in a binary mixture (2.6).
Rde	Simulates a voltammetry plot taken by a rotating-disk electrode (11.6).
Rho	Calculates partial molar volumes for ethanol-water mixtures (2.4).
rho	A file containing density data for ethanol-water mixtures (2.4).
Ro1	Calculates and plots rotational energy levels for diatomic molecules (4.5).
Ro2	Calculates and plots rotational energy levels for symmetric-top molecules (4.5).
Ro3	Calculates and plots the rotational spectrum, rotational energy levels and rotational transitions allowed by the selection rule $\Delta J = +1$ for diatomic molecules (5.1).
Rovi1	Calculates and plots rotational–vibrational energy levels for diatomic molecules (4.5).
Rovi2	Calculates and plots the rotational–vibrational spectrum, rotational–vibrational energy levels and rotational–vibrational transitions allowed by the selection rules $\Delta J = \pm 1$ for diatomic molecules (5.2).
Rovie1	Calculates and plots lines in an absorption rovibronic spectrum for a diatomic molecule (5.4).
Rrkm	Calculates and plots microscopic rate constants using the theory of Rice, Ramsperger, Kassel, and Marcus (11.4).
S&mud2	Calculates standard molar spectroscopic entropies and chemical potentials for D_2 (8.6).
S&muh2	Calculates standard molar spectroscopic entropies and chemical potentials for H_2 (8.6).
S&muhd	Calculates standard molar spectroscopic entropies and chemical potentials for HD (8.6).
Schroed1	Integrates the Schrödinger (or Schroedinger) equation, and plots wave functions, for the harmonic oscillator (4.2).

Schroed2	Integrates the Schrödinger equation, and plots wave functions, for the Morse anharmonic oscillator (4.2).
Schroed3	Integrates the Schrödinger equation, and plots wave functions, for an electron in a well (4.2).
Simha1	Calculates and plots the Simha shape factor v as a function of the axial ratio $p = b/a$ for oblate and prolate ellipsoids of revolution (7.4).
Simha2	Calculates the axial ratio $p = b/a$ given the Simha shape factor v, or vice versa, for oblate and prolate ellipsoids of revolution (7.4).
sio2	File containing powder x-ray diffraction data for $SiO_2(s)$ (silica) (6.1).
Specons	File containing special physical constants (1.1).
Splot	Calculates and plots a surface representing entropy departures for a gas at pressures and temperatures above the critical point (2.2).
Statcalc	Calculates approximate standard spectroscopic molar entropies, enthalpies, chemical potentials, and heat capacities of monatomic, diatomic, and polyatomic gaseous components (8.3).
Statg	Calculates and plots nuclear-rotational degeneracy factors for homonuclear molecules whose nuclei have spin S (8.6).
Statk	Calculates and plots spectroscopic equilibrium constants for gas-phase reactions involving monatomic, diatomic, or polyatomic (linear or nonlinear) molecules (8.5).
Step	Simulates step polymerization of monomer molecules of the A-R-B kind, and plots in a bar chart the degree of polymerization of each molecule remaining after a requested degree of polymerization (10.3).
Stokin	Demonstrates the stochastic approach to chemical kinetics (11.7).
Symtop1	Calculates and plots peaks in the parallel bands of the rotational–vibrational spectrum of a nonrigid symmetric-top molecule (5.2).
Symtop2	Calculates and plots peaks in the perpendicular bands of the rotational–vibrational spectrum of a nonrigid symmetric-top molecule (5.2).
Tcurve	Calculates and plots a curve for titration of a weak acid with a strong base (3.2).
Thermkin	Calculates and plots concentrations and temperatures for a thermokinetic oscillator (11.5).
Vanlaar	Calculates activity coefficients for components in binary mixtures according to the van Laar formula (2.7).
Vanthoff	Calculates standard Gibbs energies and equilibrium constants for reactions at high temperatures (2.8).

Vie11	Calculates and plots vibronic energy levels for diatomic molecules (4.5).
Vie12	Calculates and plots vibronic spectra for diatomic molecules (5.3).
Virial1	Calculates molar volumes for nonideal gases using the virial equation with only the second virial coefficient B included (1.2).
Virial2	Calculates molar volumes for nonideal gases using the virial equation with the second and third virial coefficients B and C included (1.2).
Waals	Solves the van der Waals nonideal gas law (1.2).
Well	Calculates energy eigenvalues for the electron-in-the-well problem (4.2).
Whitten	Calculates the sum of states $G(E)$ and the density of states $N(E)$ for a molecule's vibrational modes using the Whitten–Rabinovitch analytical method (11.4).
Xray1	Plots an x-ray diffraction pattern and calculates and plots electron densities for a known two-dimensional centrosymmetric unit cell (6.1).
Xray2	Plots an x-ray diffraction pattern and calculates and plots electron densities for a known two-dimensional noncentrosymmetric unit cell (6.1).
Xray3	Plots an x-ray diffraction pattern and calculates and plots a Patterson map for a known two-dimensional centrosymmetric unit cell (6.1).
Xray4	Plots an x-ray diffraction pattern and calculates and plots a Patterson map for a known two-dimensional noncentrosymmetric unit cell (6.1).
Xss	Calculates and plots concentrations of components participating in the reaction scheme $A + B \rightleftarrows X \rightarrow R + S$ (10.1).
Zelec	Calculates and plots electronic partition functions (8.2).
Zimm	Simulates data points and constructed lines in a Zimm plot of light-scattering data (7.5).
Zplot	Calculates and plots a surface representing compressibility factors for a gas at pressures and temperatures above the critical point (2.2).
Zrot	Calculates and plots rotational partition functions (8.2).
zrsio4	File containing powder x-ray diffraction data for $ZrSiO_4(s)$ (zircon) (6.1).
Zvib	Calculates and plots vibrational partition functions (8.2).

Appendix B
Mathematica and Physical Chemistry

A famous theoretical physicist once remarked that it is easy to get lost in mathematics. The same might be said of Mathematica, which can overwhelm the casual user with its multitude of details. Fortunately, only a small part of the full equipment provided by Mathematica (and mathematics) is needed to handle the calculational problems of physical chemistry. One of the aims of this book is to demonstrate that conclusion. As an introduction, you can read this appendix containing ten key prototype segments of Mathematica code. Adaptations of these code segments do most of the extended calculations in the programs found elsewhere in the book.

1. Equation Solving: Nondifferential Equations

The most important mathematical methods in physical chemistry calculations are those that solve equations numerically. In our problems, the Mathematica function `FindRoot` performs this task for nondifferential equations. `FindRoot` requires a statement of the equations to be solved, then a list for each variable containing a name for the variable, a starting value of each variable for `FindRoot` to use in its search for a root, and a range of values within which `FindRoot` should search.

Suppose you want to estimate the pH in a 0.1 $mol\,L^{-1}$ solution of acetic acid. The equilibrium equations to be solved simultaneously are ($Ac^- =$ acetate)

$$\frac{(H^+)(Ac^-)}{(HAc)} = 1.754 \times 10^{-5}$$

$$(HAc) + (Ac^-) = 0.1$$

$$(Ac^-) = (H^+),$$

where ( ) denotes molar concentrations with the units omitted. Using h, ac, and hac to represent the concentrations, the problem is solved with the following statement of `FindRoot` followed by code that prints the result in three digits:

```
solution =
    FindRoot[
            {h ac/hac == 1.754 10^-5,
            ac + hac == .1,
            h == ac},
            {h, .001, 0, .1},
            {ac, .001, 0, .1},
            (hac, .1, 0, .1}
    ];
    Print["pH = ", N[-Log[10, h /. solution, 3]]
```

The starting values, 0.001 for h and ac and 0.1 for hac, are estimates obtained assuming no ionization of HAc takes place, so $(HAc) = 0.1$ and

$$(H^+) = \sqrt{(1.754 \times 10^{-5})(0.1)} \cong 0.001.$$

The range prescribed for each root search, 0 to 0.1, applies the constraints that concentrations must be positive and cannot exceed the total acetate in the system. This basic strategy is elaborated in many other equilibrium calculations in Chapter 3.

2.   Equation Solving: Differential Equations

A second useful equation solver, `NDSolve`, integrates differential equations. We illustrate its use with a problem from chemical kinetics involving the reaction system

$$A \xrightarrow{1} B$$
$$B \xrightarrow{2} A.$$

Rate equations for this system are

$$\frac{d[A]}{dt} = -k_1[A] + k_2[B]$$
$$\frac{d[B]}{dt} = k_1[A] - k_2[B],$$

where [A] and [B] are molar concentrations with the units included. Suppose $k_1 = 0.2 \text{ s}^{-1}$ and $k_2 = 0.3 \text{ s}^{-1}$, and initial values of [A] and [B] are 1.0 and

$0.5\,\text{mol}\,L^{-1}$. Representing [A] and [B] with CA and CB we can integrate the above rate equations and plot the result with the code

```
k1 = .2;

k2 = .3;

CA0 = 1.;

CB0 = .5;

solution =

    NDSolve[

        {CA'[t]  ==  -k1 CA[t]  + k2 CB[t],

        CB'[t]  ==  k1 CA[t]  -  k2 CB[t],

        CA[0]  ==  CA0,

        CB[0]  ==  CB0},

      {CA, CB}, {t, 0, 10.}

    ];

Plot[CA[t] /. solution, {t, 0, 10.}]
```

The first two lines under NDSolve state the differential equations, the next two lines the initial conditions for CA and CB and the last line lists the independent variables CA and CB, and the range covered by the independent variable t in the integration. Chapter 10 contains many variations on this calculation.

3.   Matrix Manipulations

A matrix is expressed in Mathematica as a list of lists. For example,

$$m = \{\{1., \ 2.\},$$

$$\{3., \ 4.\}\},$$

represents a 2×2 matrix. Some of the problems of physical chemistry are greatly simplified by expressing them in a matrix language. Most important for us are the calculations of molecular orbital theory. The central problem is to calculate the eigenvalues and eigenvectors of the Fock matrix representing a system of molecular orbitals. The eigenvalues are orbital energies and the eigenvectors contain orbital coefficients. For example, in Hückel theory the Fock matrix for butadiene's π electrons is

$$\mathbf{F} = \begin{pmatrix} 0 & -1 & 0 & 0 \\ -1 & 0 & -1 & 0 \\ 0 & -1 & 0 & -1 \\ 0 & 0 & -1 & 0 \end{pmatrix}$$

(see Sec. 4.7 and Exercise 4-21 for additional details). Representing F with Forbital, the list of orbital energies with Eorbital, and a matrix containing the orbital coefficients with Corbital, the Mathematica code that handles this problem utilizes the function Eigensystem,

```
Forbital =
     {{0, -1., 0, 0},
      {-1., 0, -1., 0},
      {0, -1., 0, -1.},
      {0, 0, -1., 0}};
{Eorbital, Corbital} =
     Eigensystem [Forbital]//Transpose//Sort//Transpose;
Eorbital
Corbital = Transpose [Corbital]//MatrixForm
```

Eigensystem is manipulated so the orbital energies are arranged in ascending order, and the orbital coefficients, placed in the columns of Corbital, are put in the same order.

The strategy outlined is used in the programs Hueckel and Hartree. In another application the program Abc calculates principal moments of inertia as eigenvalues of the inertia matrix (see Sec. 4.5).

4.   Graphics

Many of the programs in the book display results graphically as curves plotted in color against a black background. Here is a prototype for the code that generates these displays:

```
xMin = -50.;
xMax = 50.;
yMin = -1.;
yMax = 1.;
F[x_] := Exp[-.002 x^2] Cos[x]
plotF[color_] :=
        Plot[F[x], {x, xMin, xMax},
        PlotStyle -> color,
        DisplayFunction -> Identity
        ]
```

```
Show[plotF[RGBColor[1, 0, 0]],

    Frame -> True,

    PlotRange -> {{xMin, xMax}, {yMin, yMax}},

    DefaultFont -> {"Courier-Bold", 12},

    FrameLabel -> {"x", "F[x]"},

    RotateLabel -> True,

    Background -> GrayLevel[0],

    DisplayFunction -> $DisplayFunction

]
```

The first four lines enter the x and y limits for plotting, the next line defines the function to be plotted and the next four lines a function that specifies the plot and its color. Finally, the Show function, given above with a typical collection of options, generates the display with the plotted curve in red.

5.   Data Fitting

The program Linreg performs this standard task as a hnear regression. It utilizes the function Regress from the package Statistics 'LinearRegression.' The program begins with two lists, one containing the data and the other the fitting function. For the latter we might specify

```
                fittingFunction = {1, x, x^2, x^3}
```

to fit high-temperature heat capacity data, with x representing the temperature T. The implied function for the heat capacity C_{P_m} is

$$C_{P_m}(T) = c_0 + c_1 T + c_2 T^2 + c_3 T^3.$$

Linreg fits the data to this function, calculates the coefficients c_0, c_1, c_2, and c_3, and displays the original data and the calculated function. The program is introduced in Chapter 1 and is subsequently applied many times.

6.   Simulations of Spectral Data

Chapter 5 introduces programs that simulate spectra of various kinds. Here is a prototype for the code that generates some of these spectral plots:

```
I0 = {5., 3., 10.};

a0 = {200., 150., 100.};
```

```
pf = {1000., 1700., 2000.};

nPeaks = Length[pf];

xMin = 500.;

xMax = 2500.;

yMin = 0;

yMax = 12.;

lorentz[i_, f_] :=
    IO[[i]] aO[[i]]^2/(aO[[i]]^2 + (f - pf[[i]])^2)

intensity[f_] := Sum[lorentz[i, f], {i, 1, nPeaks}]

Plot[
    Intensity[f], {f, xMin, xMax},
    PlotRange -> {{xMin, xMax}, {yMin, yMax}}
]
```

The first three lines specify data on the spectral peak amplitudes (in IO),
peak widths (in aO), and locations of the centers of the peaks, let us say as
frequencies (in pf). The fourth line calculates the number of peaks in the
spectrum. Lines 5 to 8 enter x and y limits for plotting. Next a Lorentz
function is defined which determines the shapes of the peaks. The Lorentz
contribution at the frequency f is calculated with lorentz[f], and a
plot of lorentz[f] over a range of frequencies generates the spectrum
simulation.

7.   Stochastic Models

The equations of physics are deterministic, but physical events on an atomic
or molecular level are essentially stochastic, that is, random. Usually the
deterministic models are reliable because the systems are macroscopic, and
random fluctuations are undetectable, or the theory deals in probabilities.
But occasionally stochastic modeling is of interest, as in Monte Carlo and
molecular dynamics calculations (Chapter 6), calculations of random chains
(Chapter 7), simulations of step polymerization (Chapter 10), and stochastic
formulations of rate equations (Chapter 11). These applications are diverse,
but they all rely on the generation of random numbers, in Mathematica with
the function Random. Each random number provides data for the next calculation in the program. For example, in the program Step two molecules
labeled iA and iB are chosen stochastically for a reaction from a list of n
molecules with the code,

```
While[True,
        iA = Random[Integer, {1, n}];
        iB = Random[Integer, {1, n}];
        If[iA != iB, Break[]]
    ];
```

The If function sends the program to subsequent steps if A and B are not the same molecule (A molecule cannot react with itself).

8.   Fourier Transforms

Our uses of Fourier transforms simulate applications in spectroscopy. For example, in the program Irft an interferogram of the kind generated in an infrared spectrometer is simulated and then used as the argument in the function Fourier. A plot of the result displays the spectral peaks. See Irft and Nmrft (which simulates an application to NMR spectroscopy) for more details.

9.   Symbols and Numbers

Mathematica has the remarkable ability to treat variables both symbolically and numerically. Sometimes it is expedient to include both interpretations in the same program. Here, for example, is a code fragment adapted from the program Error:

```
f[variables_] :=
                Log[x y] + (y + z)^(1/3) + Exp[x^2 z^3]
variables = {x, y, z};
numbers = {.5, .2, 1.6};
derivSymbolic[i_] := D[f[variables], variables [[i]]]
derivNumerical
    Table[
        DerivSymbolic[i] /. Thread[variables ->
        numbers],
    {i, 1, n}
    ]
```

The first line defines a (complicated) function f[variables]. Then variables lists the variables symbolically, n counts the number of variables, numbers lists numerical values of the variables, derivSymbolic[i] defines a function that calculates symbolically the partial derivative of the function with respect to the ith variable, and finally derivNumerical lists numerical values of the partial derivatives. The program Error uses this result to calculate the uncertainty in the value of the function, given a list error of uncertainties in the variables.

10. Numerical Integration

There are many recipes for accomplishing this computational chore. We use the function Nintegrate to evaluate several definite integrals with complicated integrands. For example, the program Debye calculates

$$\int_0^{T_0/\Theta_D} \frac{u^3}{e^u - 1} \, du,$$

where $T_0 = 10$ K and $\Theta_D = 100$ K, with the code

```
TO = 10.;

ThetaD = 100.;

Nintegrate[

        u^3/(Exp[u] - 1), {u, 0, thetaD/TO},

GaussPoints -> 20

    ]
```

This completes our list of useful Mathematica prototypes. Return to this list as you study the programs in the book. Try not to get lost.

Appendix C
References

M.P. Allen, and D.J. Tildesley, *Computer Simulation of Liquids*, Oxford University Press, Oxford, 1987.

S.W. Angrist, *Direct Energy Conversion*, Allyn and Bacon, Boston, 1976.

I. Barin, and O. Knacke, *Thermochemical Properties of Inorganic Substances*, Springer-Verlag, Berlin, 1979.

J.A. Beattie, and O.C. Bridgeman, *J. Am. Chem. Soc.* **50**, 3133 (1928).

P.F. Bernath, *Spectra of Atoms and Molecules*, Oxford University Press, Oxford, 1994.

M. Bodenstein, and F Boës, *Z. Phys. Chem.* **100**, 82 (1922).

C.R. Cantor, and P.R. Schimmel, *Biophysical Chemistry, Part II*, W.H. Freeman, San Francisco, 1980.

M.W. Chase, C.A. Davies, J.R. Downey, D.J. Frurip, R.A. McDonald, and A.N. Syverud, *JANAF Thermochemical Tables*, 3rd ed., American Chemical Society and American Institute of Physics, New York, 1986.

L.J. Clarke, *Surface Crystallography*, Wiley-Interscience, Chichester, England, 1985.

N. Davidson, *Statistical Mechanics*, McGraw-Hill, New York, 1962.

C.W. Davies, *Ionic Associations*, Butterworths, London, 1962.

R.M.C Dawson, D.C. Elliott, W.H. Elliott and K.M. Jones, *Data for Biochemical Research*, 3rd ed., Oxford University Press, Oxford, 1986.

J.B. Dence, and D.J. Diestler, *Intermediate Physical Chemistry*, Wiley-Interscience, New York, 1987.

J.H. Dymond, and E.B. Smith, *The Virial Coefficients of Gases*, Oxford University Press, Oxford, 1969.

J.T. Edsall, and H. Guttfreund, *Biothermodynamics*, Wiley, New York, 1983.

H. Eyring, S.H. Lin, and S.M. Lin, *Basic Chemical Kinetics*, Wiley-Interscience, New York, 1980.

J.M. Feagin, *Quantum Methods with Mathematica*, Telos/Springer, Santa Clara, CA, 1994.

W.F. Giauque, and J.O. Clayton, *J. Am. Chem. Soc.* **55** 4875 (1933).

W.F. Giauque, and T.N. Powell, *J. Am. Chem. Soc.* **61**, 1972 (1939).

W.F. Giauque, and C.C. Stephenson, *J. Am. Chem. Soc.* **60**, 1389 (1938).

R.G. Gilbert, and S.C. Smith, *Theory of Unimolecular and Recombination Reaction*, Blackwell, Oxford, 1990.

K.T. Gillen, A.M. Rulis, and R.B. Bernstein, *J. Chem. Phys.* **54**, 2831 (1971).

D.T. Gillespie, *J. Phys. Chem.* **81,** 2340 (1977).

P. Gray, and S.K. Scott, *Chemical Oscillations and Instabilities*, Oxford University Press, Oxford, 1990.

E. Hamori, *J. Chem. Ed.* **52,** 370 (1975).

Handbook of Chemistry and Physics, CRC Press, New York (1975).

N.B. Hannay, *Solid-State Chemistry*, Prentice-Hall, Englewood Cliffs, NJ, 1967.

H.S. Harned and B.B. Owen, *The Physical Chemistry of Electrolytic Solutions*, 2nd ed., Reinhold, New York, 1950.

H.G. Hecht, *Mathematics in Chemistry*, Prentice-Hall, Englewood Cliffs, NJ, 1990.

H.C. Helgeson, J.M Delany, H.W. Nesbitt, and D.K. Bird, *American Journal of Science* **278A,** 1 (1978).

B.S. Hemingway, R.A. Robie, J.R. Fisher, and W.H. Wilson, *J. Res. Geol. Survey* **5,** 797 (1977).

G. Herzberg, *Infrared and Raman Spectra of Polyatomic Molecules*, Van Nostrand, New York, 1945.

G. Herzberg, *Spectra of Diatomic Molecules*, 2nd ed., Van Nostrand, New York, 1950.

B. Hille, *Ionic Channels of Excitable Membranes*, 2nd ed., Sinauer Associates, Sunderland, MA, 1992.

K.P. Huber, and G. Herzberg, *Constants of Diatomic Molecules*, Van Nostrand, New York, 1979.

W. Kauzmann, *Kinetic Theory of Gases*, Benjamin, New York, 1965.

J. Kielland, *J. Am. Chem. Soc.* **59,** 1675 (1937).

Landoldt, Börnstein, *Numerical Data and Functional Relationships*, 6th ed., Springer-Verlag, Berlin, 1950–1980.

A.T. Larson, and R.L. Dodge, *J. Am. Chem. Soc.* **45,** 291 (1933).

I.N. Levine, *Quantum Chemistry*, 4th ed., Prentice-Hall, Englewood Cliffs, NJ, 1991.

R.D. Levine, and R.B. Bernstein, *Molecular Reaction Dynamics and Chemical Reactivity*, Oxford University Press, Oxford, 1987.

M.C. Lin, Y. He, and C.F. Melius, *Int. J. Chem. Kinet.* **24,** 1103 (1992).

L. Pauling, and E.B. Wilson, *Introduction to Quantum Mechanics*, McGraw-Hill, New York, 1935.

M.J. Pilling, and P.W. Seakins, *Reaction Kinetics*, Oxford University Press, Oxford, 1995.

K.S. Pitzer, *Thermodynamics*, 3rd ed., McGraw-Hill, New York, 1995.

J.A. Pople, and D.L. Beveridge, *Approximate Molecular Orbital Theory*, McGraw-Hill, New York, 1970.

R. Powell, *Equilibrium Thermodynamics in Petrology*, Harper and Row, London, 1978.

G. Preuner, and W. Schupp, *Z. Phys. Chem.* **68,** 157 (1910).

R.C. Reid, J.M. Prausnitz, and B.E. Poling, *The Properties of Gases and Liquids*, 4th ed., McGraw-Hill, New York, 1987.

R. Resnick, D. Halliday, and K.S. Krane, *Physics*, 4th ed., Wiley, New York, 1992.

E.G. Richards, *An Introduction to Physical Properties of Large Molecules in Solution*, Cambridge University Press, Cambridge, 1980.

R.A. Robinson, and R.H. Stokes, *Electrolyte Solutions*, 2nd ed., Butterworths, London, 1959.

J. Rosing, and E.C. Slater, *Biochim. Biophys. Acta* **267,** 275 (1972).

A. Rudin, *The Elements of Polymer Science and Engineering*, Academic, New York, 1982.

A.M. Rulis, and R.B. Bernstein, *J. Chem. Phys.* **57,** 5497 (1972).

S.I. Sandler, *Chemical and Engineering Thermodynamics*, 2nd ed., Wiley, New York, 1989.

G.C. Schatz, and S.P. Walch, *J. Chem. Phys.* **72,** 776 (1980).

S.K. Scott, *Chemical Chaos*, Oxford University Press, Oxford, 1991.

S.F. Sun, *Physical Chemistry of Macromolecules*, Wiley-Interscience, New York, 1994.

J. Zawidzki, *Z. Phys. Chem.* **35,** 129 (1907).

Index